KNOWLEDGE AND THE NORM OF ASSERTION

Knowledge and the Norm of Assertion

An Essay in Philosophical Science

John Turri

https://www.openbookpublishers.com

© 2016 John Turri

This work is licensed under a Creative Commons Attribution-NonCommercial-NoDerivs (CC BY-NC-ND 4.0). This license allows you to share, copy, distribute and transmit the work for non-commercial purposes, providing attribution is made to the author (but not in any way that suggests that he endorses you or your use of the work). Attribution should include the following information:

John Turri, *Knowledge and the Norm of Assertion: An Essay in Philosophical Science*. Cambridge, UK: Open Book Publishers, 2016. http://dx.doi.org/10.11647/OBP.0083

In order to access detailed and updated information on the license, please visit https://www.openbookpublishers.com/isbn/9781783741830#copyright

Further details about CC BY-NC-ND licenses are available at https://creativecommons.org/licenses/by-nc-nd/4.0/

All the external links were active on 22/2/2016 unless otherwise stated and have been archived via the Internet Archive Wayback Machine at https://archive.org/web

Updated digital material and resources associated with this volume are available at https://www.openbookpublishers.com/isbn/9781783741830#resources

Every effort has been made to identify and contact copyright holders and any omission or error will be corrected if notification is made to the publisher.

ISBN Paperback: 978-1-78374-183-0
ISBN Hardback: 978-1-78374-184-7
ISBN Digital (PDF): 978-1-78374-185-4
ISBN Digital ebook (epub): 978-1-78374-186-1
ISBN Digital ebook (mobi): 978-1-78374-187-8
DOI: 10.11647/OBP.0083

Cover image: Rose-Aimée Bélanger, *Les chuchoteuses* (The Gossipers, 2002), Montreal. Photo by Dan Mason. CC BY 4.0, https://www.flickr.com/photos/masondan/3681873678

All paper used by Open Book Publishers is SFI (Sustainable Forestry Initiative), PEFC (Programme for the Endorsement of Forest Certification Schemes) and Forest Stewardship Council(r)(FSC(r) certified.

Printed in the United Kingdom, United States, and Australia
by Lightning Source for Open Book Publishers (Cambridge, UK).

For Sarah,
in all her intelligence, strength, and beauty.

Love.

Contents

Acknowledgments	ix
Introduction	1
1. Evidence and Argument	7
Observational Data	7
Experimental Data	11
The Argument	16
The Explanation	16
Prefatory Remarks	18
More Challenging	20
2. Extensions and Connections	21
Know How	21
Guaranteed Knowledge	25
Knowledge Valued	29
Outstanding Questions	30
Reaching Understanding	30
Liar's Knowledge	34
3. Objections and Replies	39
Ignorant Assertions	39
Unlucky Falsehoods	39
Lucky Truths	41
Excuses, Excuses	44
Irrelevant Assessments	47

Weak Challenges	49
Pre-Theoretic Data	49
Apocryphal Paradox	50
Unbelievable Objections	52
Certain Competition	56
No Contest	59
4. Prospects and Horizons	61
What "Should"?	62
Good Enough?	65
Super Norm?	68
Requisite Truth	68
Requisite Knowledge	71
Inside and Out	72
Intuitive Connections	75
A Coincidence?	75
Why Knowledge?	77
Coda	87
References	89
Index	107

Acknowledgments

For helpful feedback on the manuscript and intellectual comradery, I thank Matthew Benton, Peter Blouw, Wesley Buckwalter, Ori Friedman, Ashley Keefner, and David Rose. Special thanks go to Angelo Turri, who commented generously and insightfully on multiple versions. Thanks to Peter Blouw for work on the index too.

My work on this book was supported by the National Endowment for the Humanities, the Social Sciences and Humanities Research Council of Canada, and an Early Researcher Award from the Ontario Ministry of Economic Development and Innovation.

I am grateful for permission to reuse material from the following publications:

Turri, J. 2012. 'Preempting Paradox', *Logos & Episteme*, 3(4): 659–62.

Turri, J. 2014. 'Knowledge and Suberogatory Assertion', *Philosophical Studies*, 167(3): 557–67.

Buckwalter, W., & Turri, J. 2014. 'Telling, Showing and Knowing: A Unified Theory of Pedagogical Norms', *Analysis*, 74(1): 16–20.

Turri, J. 2015. 'Evidence of Factive Norms of Belief and Decision', *Synthese*, 192(12): 4009–30.

Turri, A., & Turri, J. 2015. 'The Truth about Lying', *Cognition*, 138(C): 161–68.

Turri, J. 2015. 'Selfless Assertions: Some Empirical Evidence', *Synthese*, 192(4): 1221–33.

Turri, J. 2015. 'Understanding and the Norm of Explanation', *Philosophia* 43(4): 1171–75.

I dedicate this book to my daughter Sarah, whose precocious conversational acumen inspired my work on the topic. This is but one of the many ways she has inspired me and those around her. If she chooses, she will one day be a better scientist and writer than I was ever capable of.

Introduction

One road closure followed by unusually heavy traffic on the alternate route meant that we were cutting it close. Squeezing the armrest so hard that her fingernails turned white, she grimaced, "How long before we're there?" "Ten minutes," I answered, vexed that the upcoming light turned yellow. A few silent moments ensued. The uneasy thought hung over our heads like a menacing storm cloud — we might not make it to the hospital in time. Then our two year old daughter, Sarah, peeped from the back, "Daddy, how you know that?" "Know what?" "That we be there in ten minutes." "I . . . well . . . ," I faltered, blinking at the seemingly interminable red light, before continuing confidently in an attempt to reassure her, ". . . we'll get there in time for Mommy and the baby, honey, don't worry." Judging by the look on her face, my reassurance helped. But it would have helped more if I had directly answered her question in the course of reassuring her.

We did get there in time and everything went very well for both mom and baby. A few days later, as our household settled in to its new routine, I thought about the exchange with my daughter on the car ride. As our family experienced a serious and emotional situation, in the midst of all the action, excitement and concern, this wonderful little two year old, in the most natural and unselfconscious way, challenged my statement. And it worked! She asked me how I knew and, with that one innocent little question, stopped me in my conversational tracks.

Before that pregnant experience, I had been interested in assertion. In particular, I had been interested in the philosophical debate over "the norm of assertion," or under what conditions an assertion should be made. I was aware that some philosophers had suggested or argued that the default propriety of the challenge, "How do you know?" suggested that you should assert something only if you know that it is true (Unger, 1975). But in retrospect it is clear that, until that point, my interest in the debate had been "academic" in the unflattering sense of the term. After that ride to the hospital, however, my interest grew quantitatively and qualitatively. My interest was no longer academic. It was personal. I was deeply impressed by the fact that my two year old grasped the practice of assertion well enough to effectively challenge my statement in a serious situation. She implicitly understood the rules of the practice, or what anthropologists might call the "organizing principle" of this aspect of "speaking in social life" (Bauman & Sherzer 1975: 97, 110). This motivated me, in earnest, to achieve an explicit understanding of the matter — that is, of the norm of assertion and related issues. Several years later, the result is many articles and this book.

Assertion is common, unavoidable and extremely important in everyday life. Everyone reading this book already knows what assertion is, so I risk appearing patronizing by saying that assertion is an act whereby a speaker puts forward a proposition as true, and that the main vehicle for making assertions is the declarative sentence, spoken or written. There are clear antecedents of assertion in non-human animals, in the form of signaling behavior (Wheeler & Hammerschmidt 2013; Crockford, Wittig, Mundry & Zuberbuhler 2012; Bradbury & Vehrencamp 2011; Cheney & Seyfarth 1988). By asserting, we share knowledge, coordinate behavior, and advance collective inquiry. Our individual and collective well-being often depend on it. In short, assertion is fundamental to our lives as social and cognitive beings. Accordingly, assertion is of considerable interest to cognitive scientists, social scientists, and philosophers.

My focus in this book is the norm of the social practice of assertion. When *should* you make an assertion? I argue that knowledge is the norm. The basic view is not original with me. Indeed, it is an ancient idea that knowledge is the norm of assertion. For example, ancient skeptics resisted the view that we have knowledge on the grounds that we cannot properly make assertions; they also claimed that they refrained from making assertions, or even forming beliefs, because they lacked knowledge (see Turri 2012c; Sextus Empiricus, *Outlines of Pyrrhonism*). Passages from ancient religious texts convey similar sentiments. For instance, one biblical passage admonishes, "Some of you say, 'Today or tomorrow we will go to some city. We will stay there a year, do business, and make money'. But you do not know what will happen tomorrow!" (James 4:13-15, New Century Translation; cited in Benton 2012: 6). The offending parties are then told that they "should" say something different, something conditional, which presumably they do know, such as, "If things work out as planned, then we will go to a city tomorrow." We also find relevant passages from Shakespeare, who wrote, "The augurers say they know not, they cannot tell" (*Antony and Cleopatra* 4.12.2905).

In the first half of the twentieth century, philosophers began theorizing explicitly about the normative relationship between knowledge and assertion (e.g. MacIver 1938; Moore 1959), followed by explicit defenses in the later twentieth century (Unger 1975; Williamson 1996). The first decade of the twenty-first century, however, is when the view — commonly known as "the knowledge account of assertion" — began receiving serious widespread attention. In writing this book, I have built on the insights of many previous philosophers who have written on the norms of assertion, including both proponents and critics of the knowledge account.

Researchers investigating the norm of assertion agree that their project is, at least in large part, empirical. More often than not, I suspect, agreement on this point is only implicit, in no small part because contemporary philosophy bears an uneasy, disputed, and often confused relationship to empirical projects. Nevertheless, agreement on this methodological point is often explicit. As one opponent of the knowledge account put it, an adequate theory "must face the linguistic data" (Douven 2006: 45). Assertion may be a rule-governed activity but, unlike chess or cricket or legislation in modern nation states, there is no rule book. Like characters in a good Shakespearean play, we begin our project *in media res*. But our project involves a metacognitive twist that no Shakespearean character embodies: we are committed participants in a practice that inevitably acts as a medium for our attempt to achieve understanding of the practice itself. To understand the unwritten rules, we must, as it were, reverse-engineer the rulebook from the facts on the ground, from the way the practice daily unfolds. We have no choice but to begin with the available data: tendencies and habits surrounding the give and take and evaluation of assertion. We then try to discern patterns in the data. Which rules, if competently followed for the most part, would produce these patterns? Thus, while the project has a large irreducible empirical component, it also has an irreducible theoretical component too.

I approach the project in the manner of the ethologist (Tinbergen 1963; Lorenz 1974). Ethology is the scientific study of animal behavior. It begins with observation and description of the animal's behavior in natural contexts, sometimes called an "ethogram." Naturalistic observation and classification inspires hypotheses about the behavior's underlying causes. This is followed by experimentation to test the hypotheses and, ultimately, arrive at a correct explanation of the animal's behavior. Contemporary ethologists trace the ancient roots of their discipline back to Aristotle, whose work on animals anticipates many aspects of their inquiry (Menzel 2012: 609).

Good ethologists never forget, as Nobel laureate Niko Tinbergen wrote, "that naïve, unsophisticated, or intuitively guided observation may open our eyes" to new possibilities and problems (Tinbergen 1963: 417). For each of us, our everyday social interactions provide us with a continuous stream of opportunities to make relevant naturalistic observations. (It is not always easy, however, to adopt an observer's perspective on exchanges we are involved in, and some care must be taken to avoid social awkwardness due to impromptu data collection.) Previous philosophical work has contributed such observations about the human practice of assertion. In the chapters that follow, I catalogue, regiment, and expand the relevant observations into an ethogram of critical mass, in the service of formulating and, ultimately, testing the hypothesis that knowledge is the norm of assertion. I also discuss a growing body of experimental research designed to put this hypothesis to the test. To date, the knowledge account has passed every experimental test with flying colors. In the process, every serious objection to the knowledge account is addressed and previously unknown and surprising features of human moral psychology are uncovered. All the evidence points irresistibly to a single conclusion: the knowledge account is true. Finally, I also propose a hypothesis about why knowledge is the norm of assertion, informed by decades of findings from the interdisciplinary study of animal communication. Knowledge plays an important role in the evolution of stable animal communication systems generally and, I propose, its role in human communication is but a novel twist on an otherwise ancient theme.

In writing this book, I have tried to avoid what the philosopher Edward Craig once aptly called the "intellectual prejudice . . . that everything must really be frightfully complex" (Craig 1990: 4). Frightful complexity manifests itself in many ways and I have tried to avoid two of its more pernicious manifestations, length and obscurity, without sacrificing anything essential to a full appreciation of the conclusive case for the knowledge account. I should also ask the reader to keep another point firmly in mind (also partly inspired by Craig 1990: §1). My goal is not to illuminate some imaginary social practice or to prescribe some practice I think we should aspire to, but rather to illuminate the normative structure of our actual practice of assertion.

1. Evidence and Argument

In this chapter, I present the observational and experimental evidence demonstrating that knowledge is the norm of assertion. I also explain why knowledge is the norm of assertion.

Observational Data

All of us are intimately familiar with the practice of assertion. We have participated in it for as long as we can remember, as have all the people in our lives. Social observation provides a wealth of data about the ordinary give-and-take and evaluation of assertion. Introspective observation also provides further data about how certain assertions would strike us as inconsistent or odd. Taken as a whole, this set of data strongly suggests that knowledge is the norm of assertion. Everywhere we look, assertion and knowledge are linked.

Prompts. One way of prompting someone to make an assertion is to ask, "What time is it?" But an equally effective, and practically interchangeable, prompt is to ask, "Do you know what time it is?" (Turri 2010b: 458ff.). Competent speakers respond to the two questions similarly. But why would that be? Proponents of the knowledge account explain it as follows. Because knowledge is the norm of assertion, my question "Do you know what time it is?" enables you to infer that I want you to make the relevant assertion and, thus, functions as *an indirect request* for you to make the assertion. This is similar to how my question

"Can you pass the salt?" can function as an indirect request for you to pass the salt.

Abstentions When you are asked a question, even if the question has nothing to do with you or what you know, it is normally completely acceptable to respond by saying, "Sorry, I don't know" (Reynolds 2002: 140). (The exception is when it is clear that you *do* know the answer.) Suppose you are asked, "What is the conversion rate from liters to quarts?" and you respond, "Sorry, I don't know." Normally, your response will be judged perfectly acceptable. But you and your epistemic state are irrelevant to the content of the question, so why is that response any more acceptable than, say, "Sorry, Paris is the capital of France" would be? Proponents of the knowledge account explain it as follows. By saying "Sorry, I don't know," you are informing the questioner that you lack the appropriate normative standing to answer her question, which is surely relevant in the context.

Convertible. In response to a question, the statements "I don't know," "I can't tell," and "I can't say" are practically interchangeable (Turri 2011: 38). The parable of Cain and Abel contains perhaps the most famous abstention in literary history. In one translation of the story, when asked, "Where is your brother Abel," Cain answers, "I know not: Am I my brother's keeper?" (King James Version, 1611). But in another translation, Cain answers, "I cannot tell. Am I my brother's keeper?" (1599 Geneva version). Why are these locutions interchangeable? Because in ordinary speech "tell" and "say" both mean "assert," and "can" expresses the concept of permission or authority. Since knowledge is the norm of assertion, to lack authority *just is* to lack knowledge. Whence the interchangeability of all three locutions.

Challenges. When you make an assertion, even if the content of your assertion has nothing to do with you or what you know, it is normally appropriate to ask you, "How do you know that?" (Unger 1975: 263–64; Slote 1979). What explains the default propriety of this response? If knowledge is the norm of assertion, then we can explain it as follows. By making an assertion, you represent yourself as satisfying the norm of assertion; and knowledge is the norm; so the question is appropriate because it asks whether you are accurately representing yourself.

Escalation. Asking "How do you know?" is understood as implicitly challenging my authority to make an assertion. More aggressive than asking "How do you know?" is "Do you really know that?" (Williamson 2000: 252–53). More aggressive yet is "You don't know that!" or "You don't know what you're talking about." What explains this range of aggressiveness? If knowledge is the norm of assertion, then we can explain it as follows. "How do you know?" implicitly challenges my authority to assert the proposition, by asking me to demonstrate that I actually have it; "Do you really know that?" explicitly challenges my authority, by questioning whether I have it; and "You don't know that!" explicitly rejects my authority. Explicitly questioning someone's authority is more aggressive than implicitly questioning it, and explicitly rejecting someone's authority is more aggressive than explicitly questioning it.

Vindication. Suppose that you make an assertion and someone accuses you, "You're not in a position to make that claim." Responding with, "Yes I am — I know that it's true," would, if true, fully vindicate the initial assertion. Indeed, your response seems to flatly contradict the accusation. If knowledge is the norm of assertion, this is easily explained. How obtuse your accuser would seem if he answered that your response had missed the point. (Accusations made on ethical or legal grounds are different and would have to be handled differently. Such accusations are also irrelevant to my discussion here.)

Inconsistency. Assertions of the form "The match is today, but I don't know that/whether the match is today" strike us as inconsistent (MacIver 1938; Moore 1959). But their *content* is perfectly consistent, so why do they seem inconsistent? Proponents of the knowledge account explain it as follows. Knowledge is the norm of assertion, so in order to properly assert a conjunction of the form "The match is today, but I don't know that/whether the match is today," you must know each conjunct. But your knowing the first conjunct ("The match is today") would falsify the second conjunct ("I don't know that/whether the match is today"), in which case you could not possibly know the conjunction. And by asserting the conjunction, you represent yourself as knowing it, because you represent yourself as satisfying the norm of assertion (Moore 1912; Moore 1959: 173, 223; Black 1954: 54–55; Unger 1975: 253). So by asserting the conjunction, you represent yourself as knowing something that you could not possibly know, which explains the inconsistency. In a word, what you assert is inconsistent with how you represent yourself.

Oddity. It is all too common to hear things like "I can tell you that your case is still being reviewed." Consider how odd it would sound to say "Your case isn't still being reviewed, but I can tell you that it is" or "I don't know whether your case is still being reviewed, but I can tell you that it is" (Turri 2011: 39). (We get the same effect if we replace "I can tell you" in these utterances with "(what) I can say (is).") If knowledge is the norm of assertion, it is easy to explain the oddity of those assertions. The second conjunct states that I have authority to assert that your case is still under review. But the first conjunct either directly denies that I have the authority, in the case of "I don't know," or obviously entails that I lack it, in the case of "your case isn't still being reviewed."

It is worth noting that many of these observations are made in children as young as two to three years old. Developmental psychologists have documented that "know" is the most frequently used mental state verb in young children, accounting for nearly 75% of usage in some corpuses (Shatz et al. 1983: 315). The expression "I don't know" is used in young children's discourse to mean "I can't answer" (Bartsch & Wellman 1995: 42; see also Koenig, Clement, & Harris 2004), and the ability to answer a question under discussion "justifies" responding affirmatively to "Do you know the answer?" (Sodian & Wimmer 1989: 425). Young children challenge assertions made by other children and adults alike with "How you know dat?" (Bartsch & Wellman 1995: 61). Young children are also skilled at "modulating assertion," using "I think" to hedge assertions and "I know" to render them more emphatic (Shatz et al. 1983: 318–19). Children are also sensitive to the difference signaled by "I know," "I think," and "I guess" in other people's speech, which they use this to guide their actions (Moore, Bryant & Furrow 1989).

Experimental Data

Social and introspective observation are natural places to start when investigating the norm of an important social practice that we are all familiar with. And, indeed, such observations can go a long way toward clarifying the norm of assertion. The observational data reported above are quite impressive and, in my estimation, make it likely that knowledge is the norm of assertion. But some caution is warranted because social and introspective observation have some well-known limitations. As social and cognitive psychologists have shown, not infrequently we misunderstand the dynamics of social interaction and the source of our own actions and reactions (Milgram 1974: 103–04; Ross & Nisbett 2011; Lieberman 2013: 4–5).

Fortunately, the connection between knowledge and assertability can also be systemically investigated by established methods of experimental cognitive and social science. Controlled experimentation supplements introspection and social observation; it simultaneously builds on the insights they afford and overcomes their limitations, thereby increasing confidence that we have accurately identified the norm. Until very recently, inquiry into the norm of assertion has been broadly observational but not experimental. But an experimental approach is warranted because competing theories about the norm of assertion generate testable predictions, given two plausible and widely shared assumptions. The first assumption is that assertion is a social practice that competent speakers are skilled at, which is utterly uncontroversial. The second assumption is that the normative intuitions of skilled practitioners are a source of evidence about what the practice's rules are. This assumption is shared by those who appeal to competent speakers' intuitions to support theories of syntax, experimental investigations of the relationship between semantics and pragmatics, and other forms of psycholinguistic experimentation (Chomsky 1957; Noveck & Sperber 2004). It is also shared by linguistic anthropologists, who assume that a community's linguistic rules "are real for every individual member of the community, who reflects them in production, interpretation, and attitudes" (Bauman & Sherzer 1975: 113). Other things being equal, we should expect skilled practitioners to reliably identify what should and should not be done according to the rules of the practice.

Importantly, none of this implies that skilled practitioners have an explicit, articulable theory about what the rules are, or that they will answer "knowledge" if asked "what is the norm of assertion?" The assumption is not that skilled practitioners tend to be good theorists; rather, it is that patterns in their concrete, first-order intuitions and judgments manifest their skill in applying the relevant rules. Put otherwise, their intuitive reaction to cases tends to manifest their competence, resulting in detectable patterns. We can then use these patterns when theorizing about the practice's rules. The patterns will make some proposals much less likely than others, given what we expect from skilled performance.

Given these two assumptions, the normative intuitions of competent speakers are a valuable source of evidence about the norms of assertion. If that group strongly tends to judge that assertions of a certain sort should not be made, then we should conclude that those assertions probably violate the norm of assertion. Similar remarks apply when investigating the norms of other speech acts, such as questioning, commanding, guaranteeing, promising, and explanation.

The upshot of all this is that if knowledge is the norm of assertion, then full competence in the practice of assertion requires mastering the knowledge rule, and competent speakers' judgments about assertability will be guided by their commitment to the knowledge rule. Simply put, the knowledge account has testable implications. For instance, it implies that assertability judgments will be sensitive to knowledge judgments and, furthermore, to judgments about the intuitive requirements of knowledge, such as truth. Do these predictions hold up to scrutiny?

Four recent studies help shed light on the issue. One series of studies directly investigated whether the norm of assertion is, at the very least, "factive" or truth-entailing (Turri 2013b). A factive norm implies that only true assertions should be made. The studies were motivated by critics' repeated insistence that assertion's norm cannot be factive, because factive norms are highly counterintuitive and mischaracterize the practice of assertion. Instead, critics propose, the norm must be belief, or evidence, or belief supported by evidence. But the results showed that it was the critics who were guilty of mischaracterizing ordinary practice: the norm of assertion was viewed as factive. The vast majority of people judged that a well justified but false assertion should not be made, but nearly no one thought that a well justified true assertion should not be made. The results also showed that critics' favorite sort of thought experiment, intended to pump intuitions against factive accounts, can cause serious performance errors when assessing norm-violation (for more details, see the discussion of excuse validation in Chapter 3).

In another set of studies that included over nine hundred adult participants, people were asked to evaluate agents in many different situations, with different levels of evidence and with different amounts at stake (Turri & Buckwalter in press). For example, in one situation, people were asked to evaluate Jennifer, an intelligence analyst developing a file on Ivan, an elusive foreign operative. Jennifer has a source who tells her something which strongly suggests that Ivan is left-handed. Should Jennifer write in Ivan's file that he is left handed? In another situation, people were asked to evaluate Christina, a barista in charge of updating the coffee shop menu each day. To some customers with severe nut allergies, it matters whether the coffee contains pine nuts. While working on today's menu, Christina notices a persistent pattern in the supplier's shipments, which strongly suggests that the latest shipment of coffee does not contain trace amounts of pine nuts. Should Christina write on today's menu that the coffee does not contain traces of pine nuts?

In addition to answering whether the agent should inscribe some proposition, which implicates a written assertion, participants recorded judgments about many other things, including whether the proposition is true, whether the agent believes the proposition, whether the agent has good evidence for the proposition, and how important it is whether the proposition is true. Regression analysis showed that, of all these judgments, knowledge judgments had the greatest influence on judgments about whether the agent should inscribe the proposition. For example, in Christina's case, participants rated whether the coffee contains traces of pine nuts, whether Christina thinks that the coffee contains traces of pine nuts, whether Christina has good evidence for thinking that the coffee contains traces of pine nuts, how important it is whether the coffee contains traces of pine nuts, and whether Christina knows that the coffee contains traces of pine nuts. Of all these judgments, knowledge judgments had the greatest influence on judgments about whether Christina should inscribe the proposition.

A third set of studies tested the knowledge account directly, in the simplest way possible: by intervening on knowledge (Turri 2015e). That is, this study manipulated the presence or absence of knowledge by including it as an independent variable in the experimental design. This is important because if knowledge is the norm of assertion, then manipulating the presence or absence of knowledge should significantly affect people's assertability judgments. The results were overwhelmingly favorable to the knowledge account. Across a variety of scenarios, varying whether the agent knows the relevant proposition, while holding all else equal, had an astonishingly large effect on judgments of assertability. For example, consider Mallory, who manages the local farmer's market. One of her employees is interested in improving the health of his diet. The employee asks Mallory whether avocados have vitamin K. Should Mallory say that avocados have vitamin K? In one version of the story, Mallory knows that avocados have vitamin K. In the other version, she does not know. Nearly everyone who read the first story judged that Mallory should make the assertion, but nearly no one who read the second story did.

One statistic from this line of research is most impressive of all: by changing the agent's status from not knowing to knowing, the odds of judging that the agent should assert increased by a factor of nearly 350. In other words, holding all else equal, people are 35,000% more likely to judge that you should make an assertion when you know than when you do not. Knowledge judgments enormously influence assertability judgments, which is easily explained if knowledge is the norm of assertion, but hard to explain otherwise.

It might be suspected that when participants are told that an agent does not know a proposition, they infer that the agent does not believe the proposition, or does not have evidence for the proposition. For instance, when told that Mallory does not know that avocados contain vitamin K, perhaps people infer that this is because she does not believe or have evidence for the claim that avocados contain vitamin K. Accordingly, when Mallory "doesn't know," perhaps people judge that she should not assert because they think she lacks belief or evidence. If so, then the results just mentioned do not unambiguously support the knowledge account.

However, follow-up studies ruled out this alternative interpretation. The follow-up studies used slightly modified stimuli. In one follow-up study, participants in one condition were told that Mallory believes and knows the proposition, while participants on the other condition were told that Mallory believes but does not know the proposition. The difference between the two conditions was extremely large, with participants strongly disagreeing that Mallory should assert the proposition when she does not know, and strongly agreeing when she does know. In another follow-up study, participants in one condition were told that Mallory has evidence for the proposition but does not know it, while participants in the other condition were told that Mallory has evidence for the proposition and does know it. Again, the difference between the two conditions was extremely large, in exactly the same way as the follow-up study on belief. In sum, manipulating knowledge continued to have an extremely large effect on assertability judgments regardless of the presence of belief or evidence.

A fourth study addressed a similar question from a different angle (Turri, Friedman & Keefner in press). Researchers divided people into three groups. Each group read the same basic story, with one small difference. The first group was told that the agent believes a true proposition; the second group was told that the agent is certain of that same true proposition; the third group was told that the agent knows the true proposition. People then rated whether the agent should perform a variety of actions, including asserting a proposition. To illustrate, consider the following example. The water at Metro Beach was recently tested and declared unsafe for swimming. However, the health department botched the test and, as a matter of fact, the water is perfectly safe for swimming. It is a hot summer day and Alicia decides to go to Metro Beach. She examines the water and now she thinks (is certain/knows) that the water is safe for swimming. Should Alicia tell other people at the beach that the water is safe for swimming?

People who were told that Alicia *knows* agreed that she should make the assertion. By contrast, people who were told that Alicia *thinks* or *is certain* disagreed. It is worth emphasizing that this experiment held truth constant across the three conditions. That is, the comparison was not simply between knowledge, belief, and certainty. Rather, it was between knowledge, true belief, and true certainty (i.e. being certain of a proposition that is true). Whatever difference remains is attributable to knowledge specifically. And it was knowledge specifically that led people to judge that the assertion should be made.

Other experimental evidence supporting the knowledge account will be discussed in the chapters that follow. I defer discussion of these other results to a point in the presentation where they fit most naturally.

The Argument

The basic argument for the knowledge account of assertion is as simple as it is powerful: the hypothesis that knowledge is the norm of assertion is, without question and by far, the best explanation of all the available evidence. The knowledge account explains all the social, introspective, and experimental data in a simple, elegant, and unified way. It is utterly implausible that this is all just a massive string of coincidences. The sheer volume and variety of evidence that the knowledge account explains is compelling.

The Explanation

We now know that knowledge is the norm of assertion. But, it is natural to wonder, why is knowledge the norm of assertion? It is not, as some defenders of the knowledge account have suggested, "pointless to ask" this question (Williamson 2000: 267). The explanation is simple. Knowledge is the norm of assertion because the point of the practice of assertion is to transmit knowledge. In order to transmit knowledge, you must have it. This is why you should assert only if you know.

But we should not stop there. It has been argued that the norm of assertion is actually more specific and demanding than just *having* the knowledge (Turri 2011). The more demanding norm is that an assertion should *express* knowledge. The basic motivation for the more demanding view comes from reflecting on cases where someone knows that what he says is true, but his assertion does not express his knowledge. Instead, it expresses momentary confusion or his desire to cause distress or embarrassment. It seems to me that this assertion is definitely defective — it is not as it should be. But there is more than just intuition about cases here: if the point of assertion is to transmit knowledge, then assertion calls for not only the *possession* of knowledge, but also its *expression*. An assertion that does not express knowledge does not transmit knowledge. This is why you should assert only if your assertion expresses knowledge.

I accept the more specific and demanding version of the knowledge account, *the express knowledge account*, but other than the brief defense articulated in the previous paragraph, I will not dwell on it further. Instead, I will mainly discuss matters in terms of the simple knowledge account that we began with. This is purely for expository convenience: it is easier and more natural to say "knowledge is the norm of assertion" than "expressing knowledge is the norm of assertion," and it is simpler to write and read "you should assert something only if you know" than "you should assert something only if your assertion expresses knowledge."

An assertion that does not express knowledge does not transmit knowledge, so an assertion should express knowledge. Arguably this is related to what one philosopher had in mind when he wrote that the "essential character" of assertion is "knowledge-transmission" (McDowell 1998: 39–40). In the remainder of this chapter, I review some recent evidence that knowledge transmission is the point of assertion. Later, in the final section of the book's final chapter, I delve deeper into the relationship between knowledge and the practice of assertion.

Prefatory Remarks

We can and often do preface assertions with "just so you know" or other expressions implicating knowledge, such as "just so you're aware" or "just so you remember." It is perfectly natural to say, "Just so you know, our guests arrive at noon." But it is unnatural to preface assertions in ways reflective of alternative theories about the norm of assertion. According to alternative theories, the norm is belief (Bach & Harnish 1979; Bach 2008), certainty (Stanley 2008), evidence (Hill & Schechter 2007; Lackey 2007), or truth (Weiner 2005). For example, it would be odd to say, "Just so you believe this, our guests arrive at noon," "Just so you're certain, our guests arrive at noon," or "Just so you have (some) evidence, our guests arrive at noon." And it is absurd to say, "Just so it's true, our guests arrive at noon."

A related pattern emerges when we turn to prefacing questions used to prompt assertion. We can preface prompts with "just so I know" or other expressions that implicate knowledge, such as "just so I'm aware" or "just so I remember." It is natural to say, "Just so I know, do our guests arrive at noon?" But it is unnatural to preface a question with "just so I have a belief" ("just so I believe," "just so I have an opinion"), "just so it's true," or "just so I have (some) evidence." These prefaces are very unnatural, which explains why people do not use them. However, it does seem acceptable to preface prompts with "just so I'm certain." We do sometimes say things like, "Just so I'm certain, do our guests arrive at noon?"

The claims just made about the naturalness of various prefaces were recently tested (Turri in press g). Researchers divided people into two groups. Each group read a simple scenario about a married couple, Sally and Jeff. The scenario was very similar in both conditions. The main difference was that one focused on Sally giving Jeff some information (the "assert" condition), while the other focused on Jeff requesting some information from Sally (the "prompt" condition).

In the assert condition, Sally realizes that she forgot to tell Jeff that she invited guests over to watch the game. After reading the scenario, participants completed two tasks. First, they identified the most natural way for Sally to inform Jeff that their company arrives at noon. There were six options for prefacing the assertion: "just so you know," "just so you're aware," "just so you believe this," "just so you have evidence," "just so you're certain," and "just so it's true." People overwhelmingly selected the knowledge preface as most natural, and a non-trivial minority selected the awareness preface, which implies knowledge. Second, participants then rated the naturalness or unnaturalness of all six prefaces, using a 7-point scale ("very unnatural" to "very natural"). The knowledge and awareness prefaces were rated highly natural, whereas all the others were rated unnatural.

In the prompt condition, Jeff realizes that Sally did not tell him what time people were due to arrive. After reading the scenario, participants completed two tasks. First, they identified the most natural way for Jeff to prompt Sally about whether their company arrives at noon. There were six options for prefacing the prompt: "just so I know," "just so I'm aware," "just so I have a belief," "just so I have evidence," "just so I'm certain," and "just so it's true." Again, people overwhelmingly selected the knowledge preface as most natural, and a non-trivial minority selected the awareness preface, which implies knowledge. Second, participants then rated the naturalness or unnaturalness of all six prefaces, using a 7-point scale ("very unnatural" to "very natural"). The knowledge and awareness prefaces were rated highly natural. The belief, evidence and truth prefaces were rated unnatural. The certainty preface was rated natural, though significantly less natural than the knowledge preface.

If the point of assertion is knowledge transmission, then we can explain these interesting patterns in prefaces. The consistently natural prefaces indicate that the point of the assertion or prompt is no more, and no less, than achieving the point of the practice. They specify the speech act's relevance by reference to knowledge-transmission specifically. In this respect, the natural prefaces seem interestingly similar to *relevance conditionals*, such as, "If [In case] someone gets hurt, there's a first-aid kit in the closet." In a relevance conditional, the antecedent does not specify a circumstance in which the consequent is true; instead, it specifies the circumstance in which the consequent is relevant (Austin 1946; Bhatt &

Pancheva 2006). For example, someone's injury does not make it true that a first-aid kit is in the closet; instead, it specifies a circumstance where it is relevant that a first-aid kit is in the closet.

More Challenging

Earlier we noted the propriety of "How do you know?" and other challenges to an assertor's authority, which we might call *speaker-centered* or *phonocentric*. Other responses to an assertion, which we might call *listener-centered* or *audiocentric*, support the view that knowledge transmission is the point of assertion. Audiocentric responses pose a different challenge; they suggest failures of a different sort. For example, it is not uncommon to respond to an assertion with "I already know that." Rather than challenging the speaker's *authority* to make the assertion, this response challenges the assertion's *usefulness*. Another audiocentric challenge is "I don't believe you," which seems to challenge the assertion's *effectiveness*.

If knowledge transmission is the point of assertion, then we can explain the propriety of these audiocentric challenges. On the one hand, ordinarily, if someone already knows what you are telling them, then you cannot transmit your knowledge to them, because the knowledge is already in place. This explains why "I already know that" challenges an assertion's usefulness. Of course, there might be other purposes for making an assertion, as when an arresting officer tells the accused that he has the right to remain silent, or when a pupil reports progress to a teacher, but none of this spoils the present point. On the other hand, assuming that transmitting knowledge requires inducing belief, if your audience does not believe you, then you failed to transmit knowledge. This explains why "I don't believe you" challenges an assertion's effectiveness. Of course, this might indicate a failure of the listener more than of the speaker. ("You don't believe me? That's your problem.") But the challenge does have some bite.

2. Extensions and Connections

The basic argument for the knowledge account is self-contained and sufficient to compel assent in an unbiased, attentive mind. But there is yet more evidence for the knowledge account. In this chapter, I discuss six additional lines of evidence. Some are extremely well developed and constitute further compelling evidence for the knowledge account. Others are more tentative but they exhibit enough promise to be worth careful consideration.

Know How

Humans teach each other many things. We provide each other with information. Our main vehicle for transmitting information is assertion. As we leave the forest, we tell our friend headed into the forest that there is a jaguar nearby. We also teach each other skills and crafts. We show our friend how to get a jaguar to reveal its location so that he can avoid becoming its next meal. Transmitting skills is typically more intensive than transmitting information. But we are often willing to devote time and resources to doing so. This is the basis of all advanced forms of human culture and civilization.

Recall six of the observations that support the knowledge account of assertion. First, questions about what you know typically function as indirect requests to make assertions. Second, professed ignorance is a legitimate reason to avoid answering questions. Third, questions and remarks about knowledge are appropriate in light of an assertion. Fourth, such questions and remarks fall on a spectrum of aggressiveness. Fifth, citing your knowledge vindicates an assertion that is accused of illegitimacy. Sixth, certain assertions strike us as inconsistent, such as, "The match is today but I don't know that it is."

As we have already seen, these observations are explained by the fact that knowledge is the norm of assertion. And by "knowledge" we of course mean *propositional* or *declarative* knowledge — knowledge of truths or facts. But propositional knowledge is not the only sort of knowledge. There is also *procedural* knowledge, or know-how. Intriguingly, an analogous set of observations motivate a parallel hypothesis about the other main form of human pedagogy, namely, skill transmission. The parallel hypothesis is that, just as *knowing that* is the norm of information transmission, *knowing how* is the norm of skill transmission. In brief, knowing, in one form or another, is the norm of both telling and showing.

Six observations are relevant to the parallel hypothesis. First, asking whether someone knows how to do something can serve as an indirect request for instruction or a demonstration on how to do it. One way to prompt instruction is to ask, "How is this done?" but another way is to ask, "Do you know how this is done?" For example, suppose someone asks you, "Do you know how to make a campfire?" It would be perfectly natural to respond by saying, "Yes, I'll show you how." But why would that be? If knowing is the norm of showing, then the question "Do you know how this is done?" enables you to infer that this person wants you to show her and, thus, can function as an indirect request for a demonstration. This is similar to the way one's question to a bureaucrat, "Are you authorized to make an exception in this case?" can serve as an indirect request for the bureaucrat to show mercy and make an exception. Notice, furthermore, that in the case of both the campfire and the bureaucrat, it is not incompetent to respond by saying "Yes I do know how, but I will not show you" or "Yes I am authorized, but I will not make an exception in your case." Such responses might be rude but they would not exhibit misunderstanding of what such questions imply.

Second, professed inability is a legitimate reason to avoid instructing. When you are asked to provide instruction on a task, even if what you know is irrelevant to the task, it is normally appropriate to respond by saying, "Sorry, I don't know how that's done/how to do that." Suppose you are asked, "How is a shoelace tied?" and you respond, "Sorry, I don't know how to tie a shoelace." Normally your response would be judged perfectly acceptable. But you are irrelevant to the content of the question, so why is that response any more acceptable than, say, "Sorry, I get depressed when shoelaces are tied"? If knowing is the norm of showing, then by saying "I don't know how," you are informing the questioner that you lack the appropriate normative standing to show her, which is surely relevant in the context.

Third, questions and remarks about knowledge are appropriate in light of offers to instruct or attempted demonstrations. If someone offers instruction or demonstration, it is appropriate to respond, "How do you know [or: Where did you learn] how to do that?" For example, suppose that there is a group of young children, the eldest of whom is a very responsible and likeable eight-year old. The eight-year old holds up a shoe and says to the others, "Today you're going to learn how to tie a shoelace." The other children could sensibly respond by saying, "You know how to tie shoelaces?" Similarly, an adult overhearing the eight-year old's pronouncement could reasonably infer, "He knows how to tie shoelaces." Why are such responses and inferences sensible? If knowing is the norm of showing, then by offering instruction on a certain task, the eight-year old represents himself as satisfying the norm, namely, as knowing how to tie shoelaces.

Fourth, more aggressive than "How do you know how to do that?" are "Do you really know how to do that?" and, especially, "You don't know how to do that!" When the eight-year old holds up the shoe and says, "Today you're going to learn how to tie a shoelace," the other children could also legitimately respond by asking, "Do you know how to tie shoelaces?" or, if they are feeling particularly aggressive, "But you don't know how to tie shoelaces!" What explains this range of aggressiveness? If knowing is the norm of showing, we can explain it as follows. "How do you know how to do that?" implicitly challenges one's authority to provide instruction by asking how one came by the relevant know-how; "Do you know how to do that?" explicitly challenges one's authority to provide instruction by questioning whether one has it; and "You don't know how to do that!" explicitly rejects one's authority. Explicitly questioning someone's authority is more aggressive than implicitly questioning it, and explicitly rejecting someone's authority is more aggressive than explicitly questioning it.

Fifth, citing your know-how vindicates a demonstration that is accused of illegitimacy. Suppose you offer a demonstration and someone accuses you, "You're not in a position to show people how to do that." Responding with, "Yes I am — I know how to do this," would, if true, fully vindicate the demonstration. Indeed, your response seems to flatly contradict the accusation. If knowing is the norm of showing, this is easily explained. How obtuse your accuser would seem if he answered that your response had missed the point. (Accusations made on ethical or legal grounds are different and would have to be handled differently. Such accusations are also irrelevant to my discussion here.)

Sixth, certain offers strike us as inconsistent. For example, when explicitly attempting to instruct you in the acquisition of a certain skill, it would be very odd for someone to say, "I don't know how to do this, but [watch me now:] this is how it's done," or, "I don't know how this is done, but let me show you how to do it." Why do such offers seem defective? If knowing is the norm of showing, then by making the offer you represent yourself as knowing how. But then you proceed to claim that you do not know how, which explicitly contradicts the way you just represented yourself, which explains the inconsistency. The oddity here is not unlike that associated with someone (apparently sincerely) saying, "I do not know how to throw a football," while throwing a perfect spiral that hits a target thirty yards downfield. Notice also that one can qualify an offer to show by saying, "I don't know how to throw a football, but I think it's done something like this," or, "but it might be done this way." This seems analogous to the way that hedging an assertion eliminates absurdity: even though "I don't know that the match is today, but the match is today" seems absurd, "I don't know that the match is today, but I think it's today" does not.

If knowing is the norm of showing, then we can explain each of these observations in a simple, elegant, and unified way. This is strong initial evidence for the hypothesis that knowing is the norm of showing. The hypothesis is further supported by its relationship to the hypothesis that knowledge is the norm of assertion. Combining the hypotheses yields a unified theory of instructional norms: knowledge is the norm of instruction. Or, to use different terminology, knowledge is the prime pedagogical principle. The relevant form of knowledge, declarative versus procedural, depends on whether we are transmitting information or skills.

Guaranteed Knowledge

I remember very fondly a certain family vacation from several years ago. After months of planning and anticipation, the day finally came. Excitedly, we piled the children into the car and were pulling out of the driveway when my wife, Vivian, asked "Is the door locked?" "Yes, it's locked," I answered. Vivian, looking a bit concerned, began thinking aloud about a couple recent burglaries in the neighborhood. "It would be bad if we left it unlocked," she ended. I looked steadily at her and answered, "I know it's locked, Viv." Vivian was satisfied and we began our trip in earnest. (When we got back home, the door was indeed locked and the house and all our belongings were safe and sound. Lucky me.)

Just as asserting something is more emphatic than guessing, so is guaranteeing more emphatic than asserting. Someone who guarantees and turns out to be wrong is, to borrow J.L. Austin's memorable phrase, "liable to be rounded on by others" in a way that someone who merely asserts or guesses is not when they turn out wrong. One main motivation for making guarantees is to provide others with enough assurance that they are willing to proceed with a course of action in contexts where they are not satisfied with a mere assertion, as happened at the outset of my family vacation. It is a harmless oversimplification to think of guaranteeing as an especially emphatic assertion, by which you undertake heightened responsibility for the truth of the proposition guaranteed.

Many theorists have sensed that there is a close connection between *saying that you know* that something is true and *guaranteeing* that it is true (Austin 1946; Chisholm 1966; Wittgenstein 1975: §§12, 433, 575; Sellars 1975; Turri 2010a; Turri 2013a). They have noted that saying, for example, "I know that the door is locked" can be a way of guaranteeing that the door is locked. But "I know" does not mean "I guarantee," so how does saying "I know" end up being a way of guaranteeing? How does it acquire this potential?

Let us distinguish three different ways that expressions acquire this potential and then see if they can help us answer our question about "I know" and guaranteeing. But first, a word of caution: I do not want to give the impression that the differences among these categories are always hard and fast, or that, for any given expression, it is a black-and-white matter which mechanism explains its potential. There is a lot of gray area and room for improving our theoretical understanding of these issues. Nevertheless, the distinctions I am about to draw seem important and useful enough to help shed light on our main question.

First, sometimes expressions acquire their potential because of conventions that we agree on, either explicitly or implicitly. For convenience, let us call this the *conventional* mechanism. For example, there is explicit agreement that making an assertion under oath — or, for atheists, affirmation — counts as *swearing* that the assertion is true. The witness explicitly undertakes the oath and, as a result, swears by asserting. There is implicit agreement that if someone asks you to do something, then responding with "You can count on me" or "I can do that" counts as *committing* to do it. (If what you are asked to do is make a promise, then saying either of those things counts as promising.) The agreement is only implicit because no one says, "I commit to doing the things that I admit to being capable of doing." Still, such a response is *heard as* a commitment.

Second, expressions can also acquire this potential because of features specific to the conversation in which they are used, background assumptions about communicative intent, and assumptions about the speaker's goals and preferences. Let us call this the *general conversational* mechanism. For example, suppose a woman says to a man, "Let's go to the movies tonight," and the man replies, "I have a lot to prepare for a major court case scheduled early tomorrow morning." The woman made a direct proposal and the man's response would normally count as a denial. That is not because "I have a lot to prepare for a major court case scheduled early tomorrow morning" is conventionally associated with denying proposals, but rather because it is the best way to make sense of his response. To accept the proposal, all he had to say was "sure." Instead he chose to say that he had a time-consuming task to complete. Moreover, it is reasonable to suppose that declining her proposal will disappoint her, and that people prefer to disappoint others gently and politely. Conclusion: by responding the way he did, he was politely declining her proposal.

Third, expressions can also acquire this potential because of features specific to the conversation in which they are used, background assumptions about communicative intent, and facts about normative statuses like authority, permission, or entitlement. Let us call this the *normative conversational* mechanism. For example, suppose a police officer says to a motorist she just pulled over, "I'm able to let you off with a warning this time." In this way, the police officer grants mercy to the motorist. This is not because of a background assumption that police officers prefer to let speeding motorists off with a warning, as many of us have learned the hard way. There might be something conventional about the police officer's words here, but that is not all there is. Whatever the complete explanation, an important part of it is that the police officer explicitly says that she is in a position to grant mercy, that she is authorized to do so. Unless she is just being cruel and perverse, she would not mention that authority except to exercise it.

Conventional and conversational mechanisms are importantly different. With conventions, certain expressions are simply heard as a speech act of the relevant sort. Moreover, it sounds ridiculous to employ the conversational mechanism while denying that you have performed the relevant speech act. For example, if you ask, "Will you drive me to the dentist tomorrow?" it would be absurd for me to respond, "You can count on me to do it, but I'm not committing to doing it." Additionally, if I say, "You can count on me to do it," you would appear obtuse if you requested clarification: "So, just to be clear, are you committing to doing this?" Conversational mechanisms differ on all these points. The expression, "I have a lot to prepare for a major court case scheduled early tomorrow morning," is not simply heard as a denial. Something akin to an inference or calculation seems required to figure out what the speaker has done. Moreover, it is not ridiculous to deny that you have performed the relevant speech act. If the police officer said, "I'm able to let you off with a warning, but I'm not going to," she would come across as cruel but not ridiculous. Additionally, it is not obtuse to request clarification: "So you're not giving me a ticket?" (It might be unwise, though. Why push your luck?)

Let us now return to our main question, how does saying "I know that the door is locked" end up being a way of guaranteeing that the door is locked? To borrow another of Austin's memorable phrases, why does adding "I know" amount to "taking a new plunge"? Consider again the anecdote about my family vacation. Vivian asks me if the door is locked. My response: I assert that it is locked. But this does not satisfy her. She keeps the topic alive. If we are to get out of the driveway and start our vacation, I am going to have to do better. I could stop the car, get out, walk up the stairs, jiggle the handle, and then, with a blush of embarrassment, make my way back to the car. But I do not do that. Instead, I say, "I know it's locked," which satisfies Vivian and helps us on our way. But Vivian already made clear that merely asserting that it is locked would not satisfy her. So adding "I know" counts as more than just asserting — it *is* taking a new plunge. And the new depth reached is a guarantee. As Wittgenstein put it, "I know" "guarantees what is known, guarantees it as a fact" (1975: §12).

The explanation for this builds on the knowledge account of assertion. By saying "I know," you assert that you know. By asserting that you know, you represent yourself as knowing that you know. Furthermore, knowing that you know — second-order knowledge — is the norm of guaranteeing. Thus, by saying "I know," you explicitly mention that you are in a position to make a guarantee, and this is why saying "I know" has the potential to count as a guarantee. It is an instance of the normative conversational mechanism.

It makes sense that if first-order knowledge is the norm of assertion, then second-order knowledge is the norm of guaranteeing. After all, asserting and guaranteeing are "in the same line of work," namely, representing to others that certain things are true in the world. But guaranteeing is more emphatic than merely asserting; it puts more of your credibility on the line and more strongly invites others to rely on you. So guaranteeing's norm should be more demanding than assertion's. Second-order knowledge is more demanding than first-order knowledge.

Knowledge Valued

As far back as Plato's *Meno*, philosophers have wondered why knowledge is more valuable than mere true belief. If a true belief that this is the road to Larissa will get you to Larissa just as well as knowledge that this is the road to Larissa, Plato wondered, then why is knowledge better than mere true belief? This is one question about the value of knowledge.

For similar reasons, some philosophers also wonder, why is knowledge more valuable than justified true belief (Kvanvig 2003)? (Before doing any philosophy, we value knowledge. Most contemporary philosophers assume that justified true belief is necessary but not sufficient for knowledge (Gettier 1963). But only trained philosophers ever talk about "justified true belief," so it seems safe to assume that we value knowledge more than justified true belief.) A justified true belief that this is the road to Larissa will get you to Larissa just as well as knowledge will. Moreover, an analogous point holds for every status necessary but not sufficient for knowledge. Why is knowledge better than any such status? This is a second question about the value of knowledge. I do not find this question gripping, but others have.

Some philosophers also think that knowledge is better than true belief and justified true belief not only in degree but also in kind (Pritchard 2010). Why is knowledge *qualitatively* better than these other statuses? This is a third question about the value of knowledge.

Assertion is centrally important to our lives as practical, social beings. It is our primary means of communicating and receiving information needed to plan, coordinate efforts, and, more generally, live flourishing lives. So it is important for us to assert the things we should and not assert the things we should not. Whatever status that allows us to do that is valuable.

If knowledge is the norm of assertion, then we can answer all three value questions at once. First, knowledge, not true belief, licenses assertion, which explains why knowledge is better than true belief. Second, and similarly, no status insufficient for knowledge licenses assertion, which explains why knowledge is superior to any such status. Third, the difference between asserting what you should and should not is a difference in kind, which explains why knowledge is superior in kind to any status insufficient for knowledge.

Outstanding Questions

Assertion's conjugate is questioning. A question is a prompt to assertion, and a correct assertion answers the question. It would be satisfying if this reciprocity was reflected in the two speech act's respective norms. One possibility is that *knowing* is the norm of assertion while *not knowing* ("ignorance") is the norm of questioning (as suggested by Hawthorne 2004: 24). Beyond the satisfaction felt at this symmetry, there is some evidence that not knowing is the norm of questioning. For instance, if someone asks a question, the response "You already know the answer to that" challenges the question's propriety. It straightforwardly implies that the question should not have been asked. The challenge here does not seem to be based on considerations of morality, prudence, legality, etiquette, or taste. Instead, it seems to pertain to the question as such — the question *qua* question. Similarly, the responses "Now, how am I supposed to know that?" or "You know very well that I don't know the answer" suggest that it was pointless to ask the question.

Reaching Understanding

Understanding is a form of knowledge, as philosophers and scientists have recognized since as far back as Aristotle (Lipton 2004; Grimm 2006).

But not all knowledge is understanding. Beyond bare propositional knowledge of a fact or event, understanding requires knowing the answer to questions about it, such as "when?", "where?", and, more importantly, "how?", "for what?", and "why?" Allowing for the fact that understanding comes in degrees, your understanding is indexed to the number and, in some ways, detail of the relevant questions you know the answer to.

It is widely accepted that understanding is closely related to explanation (e.g. Aristotle 350BCE; Kim 1999: 11). In one sense of "explanation," one fact or event explains another by causing or otherwise producing it, but this is not the sense I am interested in here. Instead, I am interested in "explanation" as a linguistic performance, consisting of one or more assertions that answer questions about the thing being explained.

While many philosophers have offered theories about the relationship between understanding and explanation, one attractive possibility has not been explicitly identified and developed. I propose that one deep and important aspect of the relationship is normative: understanding is the norm of explanation. An explanation should express understanding. Call this *the understanding account* of explanation. This account follows from three other very plausible ideas already introduced. First, knowledge is the norm of assertion. Second, an explanation consists of one or more assertions that answer questions about a fact or event's occurrence, such as "why?" and "how?". Third, understanding consists in knowing the answer to such questions. If these three premises are correct, then the understanding account of explanation is just a special instance of the knowledge account of assertion: explanation is a special form of assertion, and understanding is the corresponding special form of knowledge.

Just like the knowledge account, the understanding account finds support in patterns surrounding the ordinary give-and-take of explanation. First, questions about understanding can function as indirect requests to provide explanations. That is, we can effectively prompt explanations by asking about understanding. For example, the question "Do you understand why/how this fire started?" is naturally understood as a request for explanation, and the response "Sure, let me explain..." is fully competent and attentive. But why would that be, if, as is true in most cases, *explanation* is irrelevant to the content of the question? If understanding is the norm of explanation, then the question "Do you understand why/how this happened?" enables us to infer that this person wants us to explain why/how it happened and, thus, can function as an indirect request for an explanation. This is similar to the way one's question to a bureaucrat, "Are you authorized to make an exception in this case?" can serve as an indirect request for the bureaucrat to show mercy and make an exception. Notice, furthermore, that in the case of the fire and the bureaucrat, it is not incompetent to response by saying "Yes I do, but I will not explain it to you" or "Yes I am authorized, but I will not make an exception in your case." Such responses might be rude but they would not exhibit misunderstanding of what such questions imply.

Second, we can appropriately abstain from offering explanations by citing lack of understanding. Suppose the topic of conversation is the recent fire and you are asked, "How did this happen?" It is perfectly acceptable to respond, "Sorry, I don't understand it myself." But you and what you understand are irrelevant to the content of the question, so why is that response any more acceptable than, say, "Sorry, I get depressed when fires occur." If understanding is the norm of explanation, then by saying "I don't understand," you inform the questioner that you lack the authority to offer an explanation, which is surely relevant in the context.

Third, questions and remarks about understanding are appropriate in light of an offer to explain events. For example, suppose someone offers to explain why the fire occurred, "Let me tell you why this happened." It is appropriate to respond, "You understand why it happened?" or, "Oh, good, I'm glad someone here understands why it happened." Why are such questions and inferences sensible? If understanding is the norm of explanation, then by offering to explain, you represent yourself as satisfying the norm, namely, as understanding. And by representing yourself this way, you make such questions and inferences sensible.

Fourth, more aggressive than "You understand why it happened?" is "But you don't understand why it happened." What explains this ordering of aggressiveness? If understanding is the norm of explanation, we can explain it as follows. "You understand why it happened?" challenges your authority to provide an explanation by questioning whether you have it, whereas "But you don't understand why it happened" explicitly rejects your authority to provide an explanation. Explicitly rejecting someone's authority is more aggressive than merely questioning whether someone has authority.

Fifth, citing your understanding vindicates an explanation that is accused of illegitimacy. Suppose the question arises, "Why did the fire occur?" and you offer an explanation. Someone levels the accusation, "You're not in a position to explain this event." Responding with, "Yes I am — I understand why it happened," would, if true, fully vindicate the explanation. Indeed, your response seems to flatly contradict the accusation. If understanding is the norm of explanation, this is easily explained. How obtuse your accuser would seem if he answered that your response had missed the point. (Accusations made on ethical or legal grounds are different and would have to be handled differently. Such accusations are also irrelevant to my discussion here.)

Sixth, certain offers strike us as inconsistent. For example, it sounds absurd to say, "I don't understand why it happened, but I can explain why it happened," or, "I don't understand how it happened, but here is how it happened…". Why do such offers seem inconsistent? If understanding is the norm of explanation, then by making the offer you represent yourself as understanding. But in the same breath you claim that you do not understand. Thus, the inconsistency results from explicitly saying that you lack the authority which you represent yourself as having.

If understanding is the norm of explanation, then we can explain all six observations in a simple, elegant, and unified way. This is good initial evidence for the hypothesis that understanding is the norm of explanation.

I suspect that the understanding account is often just below the surface in many discussions of understanding and explanation, even if no one has explicitly stated and defended it. One esteemed philosopher of science defines "explanation" as "uttering something with the intention of rendering [a fact or event] understandable," and then adds, "Such understanding I take to be a form of knowledge" (Achinstein 1983: 23). In a textbook treatment of Carl Hempel's enormously influential theory of explanation, another philosopher writes in passing, "Explanation has to do with *understanding*. So an adequate explanation of [an event] should offer an adequate understanding" of the event (Psillos 2002: 218). Passages like these suggest that philosophers of science have recognized, at least implicitly, the attractiveness of the understanding account.

Liar's Knowledge

What is it to lie? It is widely held that "there is something peculiarly odious or insulting about a lie as contrasted with other forms of deceit" (Williams 2002: 118; see also Adler 1997). Lies are assertions and cheating is insulting, so one viable hypothesis is that lying is *cheating at assertion*. To cheat is to knowingly break a rule. One possibility is that lying is asserting what you know to be false. (It might also be required that you intend to deceive your audience; I will suppress this clause in what follows.) On the current proposal, then, you follow the rule when you say what you know is true, and you cheat when you say what you know is false. Call this *the known-false account* of lying. It neatly complements the knowledge account of assertion.

Despite its elegant simplicity and pleasing symmetry, the known-false account will immediately provoke stiff resistance. In order for you to know that your statement is false, it must be false. But a standard view is that lying does not require your assertion to actually be false. Instead, you lie if you say something that you think is false. This has long been a standard view in philosophy, all the way back to at least Augustine, who wrote, "He may say a true thing and yet lie, if he thinks it to be false and utters it for true, although in reality it be so as he utters it" (Augustine 395; see also Aquinas 1273, II.II, Question 110, Article 1; Grotius 1625/2001: 258; Frege 1948: 219 n. 8; Chisholm & Feehan 1977; Bok 1978; Searle 2001: 184; Williams 2002). Social scientists adopt the same basic definition. A widely cited textbook on lie-detection says that lying is "defined solely from the perspective of the deceiver and not from the factuality of the statement. A statement is a lie if the deceiver believes what he or she says is untrue, regardless of whether the statement is in fact true or false" (Vrij 2008: 14).

Philosophers and social scientists alike motivate the standard view by appealing to "intuitions" about thought experiments (e.g., Fallis 2009). For instance, "Suppose that a suspect, who believes that his friend is hiding in his apartment, tells the police that his friend is abroad." Did the suspect lie, or did he not? "This statement is a lie [even] when, unknown to the suspect, his friend has actually fled the country" (Vrij 2008: 14).

Sometimes intuitions result from affective or pragmatic considerations, rather than manifesting competence in literally applying the term or concept in question (Sperber & Noveck 2004; Noveck & Reboul 2008; see also Chomsky 1977). On the one hand, if we disapprove of someone, we describe them in ways that reflect our disapproval (Alicke 1992). Saying that someone "lied" sounds disapproving, whereas saying that they "didn't lie" does not. On the other hand, the options available to us can influence what seems like the right answer (Guglielmo & Malle 2010). If "lied" and "didn't lie" are the only two options, then "lied" might seem right because it is closer to the truth we want to convey. If the intuitions supporting the standard view were influenced in either of these ways, then that would eliminate one main objection to the known-false account of lying.

It turns out that intuitions supporting the standard view of lying appear to be influenced in at least one of those ways. Inspired by an ingenious idea by my thirteen year old son, Angelo, he and I conducted a pair of studies in which people considered cases often thought to support of the standard view (Turri & Turri 2015). The stories all featured Jacob, whose friend Mary is being sought by the authorities. Federal agents visit Jacob and ask where Mary is. Mary is at the grocery store but Jacob thinks that Mary is at her brother's house. Jacob tells the agent that Mary is at the grocery store, so what he says is true despite his intention. In the first study, participants were asked a yes/no question:

did Jacob lie about Mary's location? A very strong majority said that he lied. In the second study, participants were offered four options and asked to select the one that best described Jacob: he tried to tell the truth and succeeded; he tried to tell the truth but failed; he tried to tell a lie and succeeded; he tried to tell a lie but failed. This time, nearly nobody said that Jacob lied and nearly everybody said that Jacob tried to lie *but failed to do so*.

It might be argued that these results do not yet undermine the standard view of lying, depending on what people mean when they say that Jacob tried to lie but failed. More specifically, it might be argued that a failed lie is still a lie, just as a failed attempt is still an attempt. To address this concern, we conducted a third study that featured different response options. Instead of asking people to distinguish between successful and failed lies, we asked them to distinguish between cases where someone *actually did lie* and *only thinks he lied*. This pair of options gives people flexibility to acknowledge the speaker's perspective while allowing them to indicate whether things actually are the way they appear to the speaker. Accordingly, we divided people into two groups. Each group read a story where Jacob intended to deceive the agent. In one version of the story, what Jacob says is false (he says that Mary is at her brother's house, but she is at the grocery story). In the other version, what Jacob says is true despite his intentions (he says that Mary is at the grocery store, and she is at the grocery store). When Jacob said something false, nearly everyone judged that he actually did lie. But when Jacob said something true, nearly everyone judged that he only thinks he lied.

These results strongly suggest that intuitions motivating the standard view are caused by having an impoverished set of possible answers in view. We should not trust those intuitions. By contrast, a good explanation of the results is that falsity is essential to lying, even though, for various reasons, people will often say that a true assertion was a lie. This conclusion is further supported by results from a subsequent study that used regression analysis and causal modeling to investigate the role of judgments about truth-value in judgments about lying (Turri & Turri under review). In this study, people's judgments about a statement's truth-value explained most of the variance in their judgments about whether the speaker lied, even when controlling for other factors, including judgments about deceptive intent.

These findings eliminate the main objection to the known-false account of lying. Of course, it does not follow that the known-false account is true. But it has two strong points in its favor: it simply and elegantly explains why lying is somehow worse than other forms of deception, and it coheres beautifully with the independently demonstrated knowledge account of assertion. In light of these facts, it deserves further careful consideration.

3. Objections and Replies

This chapter answers the main criticisms of the knowledge account of assertion.

Ignorant Assertions

Probably the most popular and persistent objection to the knowledge account is that it fumbles cases of reasonable ignorant assertions. A reasonable ignorant assertion has two features: the speaker reasonably believes that the assertion's content is true, but she does know that it is true. Critics have repeatedly discussed two types of example that supposedly fit this description.

Unlucky Falsehoods

The first type involves reasonable false assertions. In this type of case, a speaker has good evidence for believing that, say, she owns a certain type of watch. And she tells someone that she owns that type of watch. But her assertion turns out to be false despite the evidence. Perhaps the vendor mislabeled the watch so that she is wrong about what type it is, or perhaps her very reliable memory failed her on this particular occasion.

Critics of the knowledge account report having the intuition that reasonable false assertions are perfectly fine. They claim that this intuition is "obvious" and reflects ordinary practice (Hill & Schechter

2007: 109; Douven 2006: 476ff). Stronger yet, some claim that there is "no intuitive sense" in which a reasonable false assertion is improper, and that "there is no practice" of counting them as inappropriate (Douven 2006: 480; Hill & Schechter 2007: 109). If this is all on the right track, then the knowledge account cannot, without complication, "explain our intuitions about false but reasonable assertions" (Douven 2006: 478).

A series of experiments tested whether the critics have correctly described our ordinary practice of evaluating reasonable false assertions (Turri 2013b). People in these experiments considered a simple story about Maria. Maria is a watch collector who owns so many watches that she cannot keep track of them all by memory alone, so she maintains a detailed inventory of them. She knows that the inventory, although not perfect, is extremely accurate. One day someone asks Maria whether she has a 1990 Rolex Submariner in her collection. She consults the inventory and it says that she does have one. At the end of the story, one group of people was told that the inventory was right. Another group of people was told that the inventory was wrong. Everyone then answered the same question: should Maria say that she has a 1990 Rolex Submariner in her collection?

The results were absolutely clear. When the assertion would be true, virtually everyone said that Maria should make the assertion. But when the assertion would be false, the vast majority said that she should not make the assertion. This same basic pattern persisted when people were questioned in different ways. It also persisted across other differences that often can influence evaluative judgments and social cognition. For example, the pattern persisted whether the stakes were low (a "neighbor asking out of idle curiosity") or high (a "federal prosecutor asking in the course of an official investigation"). It also persisted when the stimuli were systematically switched so that the inventory said that Maria does *not* have the watch, and people had to evaluate whether Maria should make a *negative* assertion (that is, "I don't have one" as opposed to "I do have one"). When asked to explain their evaluation, a strong majority said that the statement's truth-value was more important than Maria's evidence.

One study in particular demonstrated how much subtlety and sophistication informs ordinary judgments about assertability. Instead of answering, yes or no, whether Maria should make the assertion, or rating their agreement with the statement that Maria should make the assertion, people performed a much more open-ended task of identifying what Maria should say. When the assertion would be true, the vast majority of people answered that Maria should assert that she owns the watch. But when the assertion would be false, very few people answered that way. Instead the most common response was that Maria should assert that she "probably" owns one, which, on the most natural interpretation of the case, is actually true because of Maria's evidence.

Lucky Truths

The second type of ignorant assertion discussed by critics is "Gettiered assertion." The idea here is that it is sometimes reasonable to believe true propositions that you nevertheless fail to know, due to objectionable forms of luck. These are often called "Gettiered beliefs," named after Edmund Gettier, the philosopher who sparked discussion of such examples in the mid-twentieth century (Gettier 1963; for an overview, see Turri 2012a; for some pre-1963 history of such cases, see Matilal 1986: 135–37; Chisholm 1989: 92–93). According to conventional philosophical wisdom, Gettiered beliefs fall short of knowledge. But, critics claim, intuitively there is no sense in which you should not assert Gettiered beliefs. Hence, critics argue, knowledge is not the norm of assertion (e.g. Hill & Schechter 2007; Lackey 2007; Brown 2008; Smithies 2012; Smith 2012; Coffman 2014).

"Gettier cases" come in many varieties. Here I will focus on two basic types frequently mentioned in the assertion literature. There might be no theoretically neutral way of describing the structure of these cases, but I will try to remain as theoretically neutral as possible.

On the one hand, there are "environmental threat" or "fake barn" cases (the latter label is due to Goldman 1976: 772–73, crediting Carl Ginet; see Goldman 2009: 79 n. 5). This is the most popular type of case among critics of the knowledge account. In an environmental threat case, the agent believes that something is true because she directly perceives it. If that were the end of the story, then intuitively she would know that the proposition is true. But it turns out that the agent is in an environment where her perceptual evidence could very easily have been misleading and led her to form a false belief. Intuitively, many philosophers claim, this real and very near possibility of error prevents the agent from knowing (e.g., Goldman 1976; Sosa 1991: 238–39; Neta & Rohrbaugh 2004: 401; Pritchard 2005: 161–62; Kvanvig 2008: 274). For example, suppose Sarah looks out her car window and sees a roadside barn as she drives along. Everything about Sarah and the barn is normal. But Sarah does not realize that the area she is driving through is being used as a movie set and the set designers have constructed many fake-barn façades that look just like real barns. Sarah is looking at the one real barn among all the nearby fakes. Clearly Sarah does not know that it is a barn, the critic claims, but surely Sarah should, if asked, say that it is a barn.

On the other hand, there are "explanatory disconnect" or "apparent evidence" cases (the latter label is due to Starmans & Friedman 2012). In an apparent evidence case, an agent believes a true proposition based on good but fallible evidence. If that were the end of the story, then presumably he would know that the proposition is true. But it turns out that the agent's evidence is misleading and his belief is made true by something completely unrelated to his evidence. Intuitively, the unexpected explanatory disconnect between evidence and truth prevents the agent from knowing. For example, suppose that Angelo is in the forest during deer hunting season. Two very loud, sharp bangs ring out nearby. Angelo judges that somebody is hunting deer nearby. And there is somebody hunting deer nearby. But the bangs Angelo heard were just backfire from a vehicle, and his belief is true because a camouflaged hunter is stalking a deer nearby with bow-and-arrow, silent and unseen. Clearly Angelo does not know that someone is hunting nearby, the critic claims, but surely Angelo should, if asked, say that someone is hunting nearby.

Because "explanatory disconnect" cases can be so peculiar, it is worth describing another example. Suppose that Geno's mother is completing a home improvement project and she needs a set of metric wrenches. Her old set is lost and no one can find it. So Geno goes to the hardware store, buys a new set, then puts them in the garage. But Geno did not notice that he actually bought Imperial wrenches rather than, as he thought, metric wrenches. However, there is a set of metric wrenches in the garage: his mother's old set is under some scrap metal in a garbage can where it will never be found. Clearly Geno does not know that there are metric wrenches in the garage, the critic claims, but Geno should, if asked, say that there are metric wrenches in the garage.

A series of experiments tested whether the critics have correctly described our ordinary practice of evaluating knowledge and assertion in "Gettier" cases (Turri in press b; see also Turri in press e). People in these experiments considered stories very similar to the ones described above about Sarah, Angelo and Geno. In the fake barn case, an overwhelming majority of people judged that Sarah both knew the proposition and should assert it. Indeed, the fake barn case was judged no differently than a closely matched "cheap barn" control case where there was no salient possibility of encountering a fake. This leads me to suspect that, in fake barn cases, the critics' intuitions about assertability are tracking their implicit judgments about knowledge, their ordinary competence in applying that concept. The reason it seems clear that Sarah should make the assertion is that she has knowledge. But contemporary philosophers have also been trained to say, perversely, that someone in Sarah's situation obviously lacks knowledge. This in turn causes them to misinterpret such cases as problematic for the knowledge account, even though the cases actually support it.

In the explanatory disconnect cases, judgments were more mixed. In some of them, the central tendency was to attribute both knowledge and assertability. In others, the central tendency was to deny both knowledge and assertability. Either way, the important point is that a very strong majority always kept their judgments of knowledge and assertability united, defying what critics say is the intuitive reading of such cases.

More generally the philosophical literature and lore on Gettier cases is a vast and confusing labyrinth built adventitiously over many decades. The nominal category "Gettier case" masks radical diversity in underlying causal structure. These differences are extremely important in both theory and ordinary practice — so important that it renders the nominal category, as we have inherited it, utterly useless. In the line of research just summarized, some "Gettier cases" elicited rates of knowledge attribution exceeding 80%, while others struggle to top 20%. The mere fact that something is a "Gettier case" is consistent with its being both overwhelmingly judged knowledge and overwhelmingly judged ignorance, thereby masking differences that radically affect the psychology of knowledge attributions and depriving the category of any diagnostic or predictive value (Turri, Buckwalter, & Blouw 2015; Blouw, Buckwalter, & Turri in press; Turri 2016). The lesson here is that philosophers should stop grouping into one category cases with radically different causal structures.

Excuses, Excuses

A second objection to the knowledge account is a simple argument linking blame and rule-breaking (for versions of this, see Lackey 2007: 603, 597; Douven 2006: 476–77; Hill & Schechter 2007: 109). A speaker who makes a reasonable false assertion is not thereby properly criticizable or blameworthy. A speaker who is not properly criticizable or blameworthy has probably not broken the norm of assertion. So the norm of assertion probably does not require truth. But knowledge requires truth. So the norm of assertion probably is not knowledge.

The crucial assumption here is that blamelessness is a defeasibly good indication that no rule has been broken. Is the assumption true? More importantly, does it accurately reflect the way people actually judge particular cases?

It turns out that it does not. Instead, when people consider cases of blameless rule-breaking, many prefer to describe events in a way that validates their desire to excuse. This can lead them to think and say false things about the agent's conduct. In particular, it can lead them to falsely claim that no rule has been broken. This tendency is known as *excuse validation* (Turri 2013b: Experiment 5; Turri & Blouw 2015). It is related to another tendency known as *blame validation*, which causes people to describe events in a way that validates their desire to blame (Alicke 1992; Alicke 2000; Alicke et al. 2008; see also Alicke 2008; Alicke & Rose 2010.)

Several studies compared judgments about cases of reasonable false assertion to judgments about obvious cases of blameless rule-breaking. One obvious case of blameless rule-breaking involved Brenda, who just entered a natural baking contest. Brenda just started preparing her dish in the contest. Contest rules say that only natural sugar may be used as a sweetener, so Brenda was careful to buy only sweetener clearly labeled "natural sugar." But the label on the package is wrong because there was a mix-up at the factory: an artificial sweetener that looks just like sugar was accidentally packed in a package labeled "natural sugar" without anybody noticing. Brenda is not aware that this happened and, as a result, she is actually using artificial sweetener. Obviously Brenda should not be criticized for this, and people's response to the case clearly reflects this: nearly everyone says that she should not be criticized. Equally obviously, Brenda is breaking contest rules, but people's response to the case does not clearly reflect this. When they are asked, "Did Brenda break the rules?" roughly half of people say that she did not break the rules. In other words, roughly half of people's answers contradict the plain facts of the case.

Now consider the analogous case of a reasonable false assertion. Robert recently started collecting coins. Today he made a purchase for an 1804 US silver dollar at a local coin shop. But the coin dealer cheated Robert: the coin is actually a 1904 US silver dollar that has been made to look like it says "1804" on it. Robert is not aware that the dealer did this and, as a result, he tells his dinner guests that he has an 1804 US silver dollar. Obviously Robert should not be criticized for this, and people's response to the case clearly reflects this: nearly everyone says that he should not be criticized. Equally obviously, Robert makes a false assertion, but people's response to the case does not clearly reflect this. When they are asked, "Did Robert make a false statement to his guests?"

roughly half of people say that Robert did *not* make a false statement. Again, roughly half of people's answers contradict the plain facts of the case.

Across a wide range of activities, from baking to farming to asserting to playing chess, we observe the same exact pattern of response to blameless rule-breaking: basically everyone agrees that the agent should not be blamed, and roughly half of people falsely answer that the agent did not break the rule.

Now suppose that we ask people to consider the exact same cases of blameless rule-breaking, but we change the question ever so slightly. Rather than asking whether the agent "broke the rule" or "made a false statement," instead we ask whether the agent "unintentionally broke the rule" or "unintentionally made a false statement." This slight change causes a dramatic shift — now everyone answers "yes." Everyone identifies it as blameless rule-breaking, even though half of people fail to identify it as rule-breaking. But unintentional rule-breaking *entails* rule-breaking, so how could this be?

The explanation is quite simple and, once stated, can seem completely obvious. People answer "no" to the original question because they want to avoid indirectly blaming a blameless agent. A factually accurate answer — "yes, she broke the rules" — could easily seem unfair and many people prefer to avoid giving that impression. The adverb "unintentionally" is often used to indicate that the agent should not be blamed for a bad outcome. So the modified question — "yes, she *unintentionally* broke the rules" — liberates people to answer accurately; it does not force them to choose between answering accurately and avoiding unfairness. Instead, by agreeing that the agent unintentionally broke the rules, people can simultaneously accurately identify the rule-breaking and excuse it.

Excuse validation is a very robust tendency. As already mentioned, it occurs when evaluating a wide range of activities. We observe it in both women and men. It persists when the consequences of rule-breaking are trivial and when they are momentous. For example, in one study less than half of people said that an agent broke the rules when the result was that a database would have to be updated manually. In a closely matched condition, less than half of people said that the agent broke the rules when the result was that the nation goes to war! Excuse validation also occurs when people evaluate other people's statements about blameless rule-breaking, rather than judging it directly for themselves.

Taken together these findings completely undermine the attempt to use cases of reasonable false assertion against the knowledge account. A predictable proportion of people react to blameless rule-breaking by engaging in excuse validation: they literally deny that a rule was broken, even when it obviously was broken. Just as Brenda blamelessly broke the baking contest's rule by using artificial sweetener, so too did Robert blamelessly break assertion's rule by making a false statement. The critic's intuitions here are simply excuse validation in action.

John Stuart Mill once wrote of moral judgment, "We do not call anything wrong unless we mean to imply that a person ought to be punished in some way or other for doing it" (Mill 1863/1979: ch. 5). This insightful observation is but part of a much larger picture: many of us are unwilling to identify even trivial, non-moral instances of blameless rule-breaking as rule-breaking. And many of us are willing to contradict others who accurately identify blameless rule-breaking as rule-breaking.

Irrelevant Assessments

In keeping with prior theoretical work on assertion, much of the experimental work discussed here assumes that the "should" of assertability differs from other familiar sources of normativity, such as morality, rationality, politeness, or legality. That is, assessments of assertability ordinarily do not reduce to assessments of the assertion's morality, rationality, etiquette, or legality. To illustrate this assumption with an analogy, consider a chess match. The goal of chess is to checkmate your opponent. The rules of chess allow rooks to move along an unobstructed vertical or horizontal path. If you can checkmate an opponent by moving a rook along an unobstructed vertical path, then there is a clear sense in which you should make that move. But if your opponent is a child who would be utterly devastated by the defeat or a violent mobster who will react violently to a loss, then there is also a clear sense in which you should not make the move. In these ways, the normativity distinctive of chess differs from the normativity of morality or practical rationality. Similarly, experimental research on assertion has assumed that there is a distinctive "should" of assertability.

Contrary to that assumption, some theorists have worried that patterns favoring factive accounts "actually track moral considerations rather than those that are proper to assertion" (Pagin 2015: 22). Similarly, one might worry that the attributions are tracking assessments of rationality, etiquette, or legality.

These worries have been directly tested experimentally (Turri under review). People were divided into groups and read a brief story. Everyone read the same basic story in which an agent has evidence for a proposition and is asked whether it is true. In one version of the story, the proposition is true; in another version, the proposition is false despite the evidence. Researchers also varied how much was at stake for the agent. For instance, she might be having an idle conversation with a neighbor (lower stakes), or under question by a federal prosecutor (higher stakes). After reading the story, participants rated whether the agent should make the assertion. Participants also rated the assertion's morality, rationality, etiquette, and legality, in addition to its truth-value and how serious the situation was for the speaker. Researchers then used regression analysis to statistically analyze which of these judgments and other variables predicted assertability attributions.

The results ruled out the worries and provide further strong evidence that assertion has a factive norm. Even when controlling for all the other factors' influence, evaluations of truth-value significantly predicted assertability attributions. Indeed, evaluations of truth value were the *strongest* predictor. This occurred when the stakes were lower and when they were higher. No other quality significantly predicted assertability attributions in both stakes conditions. When the stakes were lower, evaluations of etiquette also contributed significantly to assertability attributions. When the stakes were higher, evaluations of rationality and legality also contributed significantly to assertability attributions. Regardless of stakes, evaluations of morality and the seriousness of the situation did not predict assertability attributions. Assertability attributions were also unaffected by participant gender or age.

Weak Challenges

The observational data supporting the knowledge account include appropriate challenges to assertions. When someone makes an assertion, it is normally perfectly appropriate to ask them, "How do you know that?" or, more aggressively, to say, "You don't know that." But it is also perfectly appropriate to say, "That is not true," "All the evidence suggests otherwise," or, "You don't believe that." Don't these latter challenges support weaker accounts of assertion's norm, namely, a truth account, an evidence account, or a belief account (Kvanvig 2009)?

Taken in isolation, the propriety of these challenges does provide some evidence for the alternative accounts mentioned. But it does not favor these alternative accounts over the knowledge account because the knowledge account explains their propriety very well. We have theoretical and empirical evidence that, on the ordinary conception of knowledge, knowledge requires truth, belief, and not believing what goes strongly against the evidence (Buckwalter 2014; Starmans & Friedman 2012; Buckwalter, Rose & Turri 2015; Turri, Buckwalter & Blouw 2015; Turri & Buckwalter in press). So to question whether an assertion is true, whether the speaker believes what he is saying, or whether it goes strongly against the evidence is, by implication, to question whether an assertion expresses knowledge. The knowledge account, therefore, easily explains the relevance of these weaker challenges. More generally, the knowledge account easily explains the propriety of any challenge featuring an intuitively plausible requirement of knowledge.

Pre-Theoretic Data

Some critics argue that some of the observational data we have been discussing are less "pre-theoretic" and more tendentious than I have supposed. For instance, consider again the challenge, "That's not true," said in response to an assertion. Advocates of the knowledge account assume that the challenge constitutes a criticism of the assertion, as opposed to merely forcing the speaker into a position where he must either defend his assertion or retract it. But is it intuitively clear that the challenge really does constitute a criticism of the assertion? Critics claim "not to share such intuitions" and they suspect that the supposed datapoint is actually a "theory-laden intuition" (Rescorla 2009: 123). The most substantive, genuinely "pre-theoretic" datapoint we should allow, they argue, is that when someone challenges your assertion, you must either defend its truth or retract it. This falls short of the claim that a false assertion violates the norm of assertion (Rescorla 2009: 125).

I note two points in response to this line of reasoning. First, it is certainly true that multiple explanations are possible for any particular observation. But one-off explanations are cheap and rival hypotheses must be judged by how well they explain the entire range of data. It remains to be seen whether the proposed "defend or retract" account can well explain other relevant phenomena, let alone the entire range of data on the table. Second, the available evidence fully addresses the speculative worry that the intuitions in question are "theory-laden." In multiple studies, the vast majority of people judged that false assertions should not be made. This is certainly not due to some theoretical commitment that these people all happened to share. Moreover, in many cases these judgments were prospective. No assertion had yet been made, let alone challenged. While a "defend or retract" norm is logically consistent with this — it is possible that people were quickly and implicitly evaluating a counterfactual situation in which the speaker is challenged and cannot defend himself — the fit is strained and ad hoc. But why settle for that when the knowledge account is a perfect fit?

Apocryphal Paradox

Some have argued that a certain version of the knowledge account entails a paradox and so must be false (Pelling 2011, 2012). The version in question says that knowledge is not only necessary for assertability, but also sufficient. That is, knowledge is both necessary and sufficient to license assertion. For the sake of argument, grant that this stronger ("biconditional") version of the knowledge account is preferable.

We can represent the argument as proceeding in three separate stages. First, we are asked to consider an isolated utterance of a sentence named "A1": "This assertion is improper." Second, we are told that an utterance of A1 causes serious trouble for the hypothesis that truth is both necessary and sufficient to license assertion. What serious trouble? Suppose that I say, "This assertion is improper." If my assertion is true, then it is improper. If my assertion is false, then it is improper. Either way, we are told, it is a counterexample to the truth account. This creates "a self-referential paradox for the truth account." Third, we are told that the paradox extends to afflict the knowledge account if it is possible to know two things: on the one hand, that the knowledge account is true; on the other hand, that if the knowledge account is true, then an utterance of A1 is true. A bit of further conditional reasoning leads to the paradoxical conclusion that if the knowledge account is true, then it is true *both* that I can know that A1 is true, *and* that I cannot know that A1 is true. Since it implies a contradiction, the knowledge account cannot be true.

I note two points in response to this argument. The first point is that someone who responds to the knowledge account this way has misunderstood the nature of the project. Even supposing that the argument works flawlessly, it is misguided to conclude that knowledge is not the norm of assertion. For there is no reason to suppose that, when combined with various other assumptions, a social practice's rules will not have contradictory implications. To illustrate the point, suppose we are discussing the official rulebook of a legislative chamber (or a chess club, baseball league, baking contest, etc.). Now I proceed to prove that, when combined with assumptions about weird self-referential acts that no one ever actually performs, the rules imply a contradiction. Would it follow that these rules are not the legislative rules after all? Of course not! The entire exercise was meant to show that *these rules* have some paradoxical implication. If this in turn implied that they were not the legislative rules, then we would have, paradoxically, failed to show that the legislative rules have a paradoxical implication. Similarly, returning to the knowledge account, even if it did paradoxically imply an inconsistency, it would not follow that knowledge is not the norm of assertion.

The second response is that the argument does not work flawlessly. In fact, once identified, its principal assumption appears highly dubious. The argument assumes that an isolated utterance of "This assertion is improper" counts as asserting some definite proposition. But no reason is offered in favor of this crucial assumption and there is reason to doubt it. If someone weirdly uttered the isolated sentence "This command is improper" or "Obey this command," they would not be offering a command. The most natural reaction to such an utterance is to wonder, "What command are you talking about?" Similarly, if someone weirdly uttered the sentence "This question is improper" or "Is this question improper?" it is, at the very least, unclear that they would be asking an actual question. The natural reaction is to wonder, "What question are you talking about?" Analogous points can be made about weird utterances involving other speech acts such as, "This guess is improper," "This hypothesis is improper," "This announcement is improper," and so on. My reaction to a de-contextualized utterance of "This assertion is improper" follows precisely that pattern. I am left wondering, "What assertion are you talking about?"

Of course, we can easily imagine contexts in which some particular assertion is the topic of conversation — perhaps some particularly outrageous statement by a politician, comedian, or bigot — in which case saying "This assertion is improper" would make good sense. But that is because in that context we naturally interpret the noun phrase "this assertion" as referring to a salient pre-existing assertion. But the argument against the knowledge account assumes a radically different understanding of the phrase "this assertion." In order for the argument to get off the ground, that phrase must be understood self-referentially and, furthermore, in such a way as to imply a contradiction. But it is highly doubtful that the phrase ever must be understood that way.

Unbelievable Objections

Some critics argue that knowledge is not the norm of assertion because knowledge requires belief, whereas assertability does not require belief (Lackey 2007). The argument is, as usual, defended almost entirely by appealing to intuitions about thought experiments. The thought experiments feature what are called "selfless assertions." A "selfless assertion" is supposedly an assertion that has two crucial features. First, it is an assertion that, intuitively, the agent should make. Second, we naturally interpret the agent as neither believing nor, as a result, knowing the proposition asserted. By this point, the attentive reader will immediately question the dependability of these intuitions. And such skepticism would be very well placed.

The most widely discussed example of "selfless assertion" features Sebastian, a well-respected pediatrician and researcher who has extensively studied childhood vaccines (Lackey 2007: 599). Sebastian "recognizes and appreciates that all the scientific evidence shows that there is absolutely no connection between vaccines and autism." But Sebastian's own eighteen-month-old daughter was recently diagnosed with autism shortly after receiving one of her vaccines. The emotional trauma of his daughter's diagnosis causes Sebastian to begin doubting his previous views about vaccines and autism, and he is aware that this is the source of his doubt. Moreover, he still recognizes that the evidence shows that there is no link. So when a baby's parents ask Sebastian about the rumors of a link, he tells them, "There is no connection between vaccines and autism."

Critics claim that two things are obvious about Sebastian. First, he does not believe that there is no link between vaccines and autism. Second, he should tell the parents that there is no link. So Sebastian should assert what he does not believe. Assuming that knowledge requires belief, it follows that Sebastian should assert what he does not know. Critics conclude that the knowledge account faces a "fundamental difficulty" (Pritchard 2014: 160; see also Wright 2014: 255).

Some researchers have responded to cases like Sebastian's by proposing that, on the most natural interpretation of the case, he *does* believe that there is no link (Turri 2014c). Who is right? In order to answer that question, we need better evidence on how the case actually is most naturally understood. Do people judge that Sebastian should make the assertion? Do people judge that Sebastian believes the claim in question? Do people judge that Sebastian knows the claim in question?

A recent study investigated these questions experimentally (Turri 2015c). The results confirmed that it is definitely intuitive that Sebastian should assert that there is no link between vaccines and autism: over 80% agreed that Sebastian should make the assertion. However, it is also definitely intuitive that Sebastian both believes and knows that there is no link: nearly 90% attributed belief and knowledge. These results completely contradict the critic's interpretation of the case and, ironically, end up providing further confirmation of the knowledge account.

Another example of "selfless assertion" features Stella, a "devoutly" religious "creationist teacher" who teaches science to fourth-graders (Lackey 2007: 599). Stella's "deep faith" includes "a belief in the truth of creationism and, accordingly, the falsity of evolutionary theory." Nevertheless, Stella "fully recognizes" the "overwhelming scientific evidence against creationism and in favor of evolutionary theory." This leads Stella to tell her students, "Modern humans evolved from more ape-like ancestors called *hominids*." Should Stella make this assertion? Does she believe that humans evolved? Does she know that humans evolved? When this case was tested, people overwhelmingly agreed that she should make the assertion. However, they also overwhelmingly agreed that she believes and knows that humans evolved. Again the results completely contradict the critic's interpretation of the case and provide further confirmation of the knowledge account.

Critics have offered other examples of "selfless assertion." But they are ill suited to test intuitions about assertability. They involve provocative, even incendiary, subject matter that can potentially interfere with people's judgment. For instance, one case involves a "racist juror" sitting in judgment of an innocent black man accused of interracial sexual assault. The experiments discussed above focused on less provocative but still emotionally and morally charged examples. The examples of Sebastian and Stella involve socially controversial issues: the safety of vaccines and the antagonism between creationism and evolutionary theory. The stories also raise the prospect of harming innocent babies and children by threatening their physical health or intellectual well-being. It is not mere speculation that all this will trigger strong moral feelings. In the very same study discussed above, people also said it would be highly immoral for Stella to not make the relevant assertion. Moreover religious belief has a privileged social status in Western culture, so many people might feel uncomfortable explicitly attributing beliefs that conflict with someone's avowed religious faith.

Aside from involving highly emotionally charged themes, all the cases critics have discussed are long, complicated, and confusing. They are confusing because they send mixed signals about the agent's state of mind. For example, the agent is described as "fully recognizing" that there is an "overwhelming amount of scientific" evidence in favor of a certain proposition, but in the same paragraph it is explicitly stipulated that the agent "neither believes nor knows" the proposition. In other cases the agent is described as experiencing a cognitive roller-coaster, first knowing, then doubting, then "recognizing" that the doubt was irrational, followed by asserting the proposition in question.

In general, theoretical debate is not well served by focusing on complicated, confusing, and provocative cases. They introduce irrelevant factors that could easily cause performance errors or otherwise degrade social cognition. And yet, despite all of that, when tested these cases produced results fully consistent with the knowledge account.

But the defects of particular cases are not the fundamental issue. A deeper problem lurks here: thought experiments intended to probe for mental state attributions should not conflict with basic principles that guide social cognition. Previous work on social cognition shows that assertion is a powerful cue to belief attribution. Indeed, assertion can sometimes be a stronger cue to belief attribution than even a robust and consistent profile of non-verbal behavior (Rose, Buckwalter & Turri 2014). And work in developmental psychology shows that even very young children operate with a default assumption that people believe what they say (Roth & Leslie 1991; see also Nichols & Stich 2003). Even if critics devise simpler, coherent, more mundane cases of "selfless assertion," we cannot magically stipulate away our tendency to interpret people as believing what they say. If thought experimentation is worth doing, it is worth doing well.

Certain Competition

As we have seen, critics have tried to produce counterexamples to the knowledge account. The counterexamples are often interpreted as motivating weaker norms of assertion, such as belief or justification. But these counterexamples have been carefully studied and, one by one, they have all been effectively dismissed. However, a different objection to the knowledge account does not proceed by trying to pump intuitions about alleged counterexamples. Instead, it tries to identify data that the knowledge account might not explain so well.

One observation that the knowledge account well explains is the default propriety of many challenges to assertion. For instance, when I make an assertion, even if the content of the assertion has nothing to do with me or what I know, it is still normally appropriate to ask, "How do you know that?" If knowledge is the norm of assertion, then we can explain the propriety of this question by pointing out that by making the assertion I represent myself as knowing. However, it also seems appropriate to ask, "Are you certain?" or, "How can you be sure?" If we assume, as many do, that knowledge does not require certainty, then the knowledge account cannot as simply explain the propriety of this latter challenge. Some take this to motivate the certainty account: you should assert a proposition only if you are certain that it is true (Stanley 2008).

Some have proposed explanations of the "certainty" challenge that are consistent with the knowledge account. For instance, some have suggested that to be certain is, roughly, to know that you know. The propriety of the "certainty" challenge could then be explained as follows: by making an assertion you represent yourself as knowing, and the "certainty" challenge is appropriate because it asks you whether you have accurately represented yourself (Turri 2010b).

Alongside this explanation, it has been proposed that data on how we *prompt* assertion show that assertability is more closely connected to knowledge than certainty. For example, we naturally prompt assertion by asking, "What time is it?" Equally naturally, we can prompt assertion by asking, "Do you know what time it is?" Competent speakers respond to these similarly. The knowledge account can explain this on the grounds that we prompt assertion by asking whether you satisfy the norm of assertion, just as we can make a request by asking whether someone is in a position to grant the request — for example, one might ask an officious bureaucrat, "Are you authorized to make an exception in this case?" By contrast, we do not naturally prompt assertion by asking, "Are you certain (about) what time it is?" Questions about certainty typically become appropriate only after an assertion has been made. Aside from this, proponents of the certainty account have yet to address the very large amount of observational and experimental data discussed above in Chapter 1.

There is also a more direct test of the competing proposals. If assertion is more closely connected to knowledge than certainty, then this will have detectable behavioral consequences. In particular, it implies that people will be more willing to attribute *assertability without certainty* than *assertability without knowledge*. The matter has been tested (Turri in press c). People read a story about Angelo very similar to one discussed above. Angelo is camping with his daughter in a wooden cabin at the edge of the forest. As they settle in to sleep for the night, the daughter has her headphones on and Angelo is reading near the window. Angelo hears two very loud, sharp bangs ring out in the forest behind the cabin. It is deer-hunting season. Angelo's daughter takes off her headphones and asks, "Dad, what's going on? Is somebody hunting deer nearby?" After reading the story, one group of people was asked to evaluate assertability in relation to certainty, while another group was asked to evaluate assertability in relation to knowledge.

The primary question is how frequently people were willing to unlink certainty and assertability, on the one hand, and knowledge and assertability, on the other. A *unified response* keeps the epistemic status and assertability together. For knowledge, a unified response either attributes both knowledge and assertability, or denies both knowledge and assertability. For certainty, a unified response either attributes both certainty and assertability, or denies both certainty and assertability. A *disunified response* is simply the opposite of a unified one. The results were clear. The vast majority of people offered a unified response for knowledge, whereas only half of people offered a unified response for certainty. In fact, the odds of someone offering a disunified response was over four times greater for certainty than for knowledge. Assertability is more closely linked to knowledge than to certainty.

Related experimental findings were discussed in Chapter 1. To briefly reiterate — and focusing only on the results relevant to comparing knowledge and certainty — people were divided into groups (Turri, Friedman & Keefner in press). Everyone read the same basic story, in which the key proposition is true. But there was one small difference. In one version of the story, the agent is *certain* that the proposition is true. In the other version, the agent *knows* that it is true. People then rated whether the agent should assert the proposition. People who were told that the agent *knows* agreed that she should make the assertion. By contrast, people who were told that the agent is *certain* disagreed. This difference emerged even though truth was held constant across the conditions. Again, assertability is more closely linked to knowledge than to certainty.

Philosophers have said and assumed many things about the relationship between knowledge and certainty (for example, Descartes 1641/2006; Unger 1975; Wittgenstein 1975; Moore 1959; Klein 1981; Chisholm 1989). But very little is known about how these categories are related in ordinary social cognition. Some notable theorists have argued that knowledge requires being rightfully sure of a proposition (Ayer 1956: 34), though nowadays there seems to be wide agreement among professional philosophers that knowledge does not require certainty. But much recent empirical work has shown that professional philosophers often have, or at least report having, idiosyncratic and stylized intuitions about knowledge and related matters. Moreover, philosophers often seem unaware that their intuitions and assumptions deviate substantially from deep patterns in ordinary social cognition. It would be valuable to investigate how knowledge and certainty are related in ordinary social cognition. The results could then inform theorizing about the norms of assertion. In particular, they could reveal a form of certainty that is equivalent to, or required for, knowledge, as it is ordinarily understood. I would not be surprised if that turned out to be true. If it does, then the knowledge and certainty accounts are not necessarily competitors after all.

No Contest

A few years ago, when I first considered writing a book defending the knowledge account, I imagined that it would include a chapter or two dedicated to evaluating rival accounts in detail. That is because, in the past, critics tried to argue that alternative accounts could explain all the evidence as well as the knowledge account did. Those days are now gone, however, because recently the quantity and variety of evidence has grown exponentially. Indeed, so much additional evidence has accumulated that rival accounts are essentially forced back to the starting line. Additionally, some evidence that rivals spent considerable time trying to explain, such as assertions about losing lottery tickets, never impressed me and is entirely absent from the present discussion. Nowadays the knowledge account has no rival. Instead of a chapter or two, all that is left is the previous section on certainty and this lonely paragraph. Would-be critics have their work cut out for them — provided that they do not mind working in vain.

4. Prospects and Horizons

Over the past decade, the development of the literature on norms of assertion has been extraordinary. It has been one of the most lively and fruitful areas of philosophical inquiry during this time. And, as should be clear from the preceding chapters, it has paid enormous dividends. We now know that there is a deep normative connection between knowledge and assertion. The best way to understand this connection is that knowledge is the norm of assertion. That one simple idea packs immense explanatory punch. The amount and variety of evidence that it explains is astounding and, as far as I am aware, unprecedented in philosophy's recent history. It is a hard-won discovery that illustrates philosophy's value, exemplifies genuine philosophical progress, and is something the discipline can be proud of.

The knowledge account reveals something deep and important about an absolutely central aspect of our lives as social beings. It identifies the core evaluative principle of our information-sharing practices and, in the process, illuminates a central plank in normative social cognition. I consider it to be one of the most significant contributions that contemporary philosophy has made to our understanding of the human condition — though I acknowledge, of course, the possibility of reasonable disagreement on this last point. But however the grand bookkeeping works out, further progress in this area will come not by asking whether knowledge is the norm of assertion. That would be pointless because we already know that it is. Even worse would be to continue the misguided hunt for alleged counterexamples to the knowledge account. Instead, further work should seek, on the one hand, to sharpen and extend our understanding of the normative relationship between knowledge and assertion and, on the other, what clues this might give us to other important questions.

What "Should"?

Many researchers working on the norm of assertion have accepted, often implicitly, several assumptions. They assume that there is a *unique* norm of assertion. They assume that the norm is *rule-like* because it says that you should make an assertion only under certain conditions, as opposed to, say, stipulating when one assertion is better than another. They assume that the rule is *deontological* because the "should" expresses the concept of permission; you have permission or authority to assert only under certain conditions, and to do otherwise is impermissible. They assume that the norm imposes a *perfect* requirement or standard, one that applies strictly to each and every assertion. They assume that the norm is *concurrently restrictive* because the condition must be satisfied prior to or concurrent with the assertion, rather than afterward. They assume that the norm is *discretionary* because it leaves it to your discretion whether to exercise your authority to assert; it does not obligate you to make any assertion. They assume that the norm is *constitutive* because it constitutes, organizes, or sustains the social practice of assertion, similar to how the rules of a government are essential to making it the sort of government it is, or how the rules of chivalry were essential to making it the cultural practice it was. Finally, they assume that the norm is also *individuating* because it distinguishes assertion from related speech acts such as guaranteeing or guessing, which are governed by different norms.

We could propose a theory of the norm of assertion that rejected any or all of these assumptions. We might conjecture that assertion has no norm or multiple norms. We might conjecture that the norm does not say when an assertion should be made but, rather, only when one assertion is better than another. We might conjecture that the "should" expresses not the concept of permission but, rather, goodness; on this view, asserting what you do not know is not impermissible but bad. We might conjecture that the norm imposes only an imperfect requirement; for example, perhaps it requires only that most of your assertions express knowledge, or that the central tendency in your overall pattern of assertions is the expression of knowledge. We might conjecture that the norm only requires you to do certain things after your assertion has been challenged rather than meeting some standard at the time you assert. We might conjecture that the norm obligates you to make assertions under certain conditions. We might conjecture that the norm does not help constitute the practice of assertion but instead is a norm of morality or prudence. Finally, we might conjecture that the norm does not individuate assertion but instead is common to many speech acts.

The evidence discussed in this book unquestionably demonstrates a deep normative relationship between assertion and knowledge. But it does not necessitate a unique interpretation of that relationship. In other words, although it is clear that an assertion should express knowledge, we do not yet fully understand this "should." It would be counterproductive and misleading to claim false precision on the issue at this point. For instance, nothing in the evidence requires that knowledge is the unique norm of assertion or that the norm is purely discretionary. Nevertheless, the available evidence does help us begin narrowing things down. It does so in at least five ways.

First, it supports the assumption that the knowledge norm is rule-like. People responded consistently to questions about what "should" or "should not" be asserted, which is rule-like. Second, it supports the assumption that "should" expresses a status more like permission than goodness. People react to reasonable false assertions by engaging in excuse validation and, as far as we know, excuse validation seems to occur when someone does something forbidden. Nevertheless, I acknowledge that further investigation could reveal greater complexity on this point (more on this in this chapter's next section). Third, it supports the assumption that the norm imposes a perfect requirement. In response to the challenge, "You don't know that," it would be silly to respond, "Maybe not, but that's no problem because I know most of the things I say." But if the norm were imperfect, then such a response arguably should sound fine. Consider the paradigm case of an imperfect duty: charitable giving. If someone challenges you, "You didn't give to charity this week," it would be perfect sensible to reply, "Maybe not, but that's no problem because I give most weeks." Nevertheless, again, I do not think that this conclusively settles the issue. Further investigation could reveal evidence best explained by a stringent but still imperfect norm.

Fourth, the evidence suggests that the norm is concurrently restrictive. Many of the experimental studies asked people to evaluate assertions prospectively. A situation was described in which an agent had certain evidence and was asked a certain question. The agent had not made an assertion yet and, obviously, no one had challenged her assertion. Nevertheless, people's assertability judgments were powerfully influenced by the presence of both truth and knowledge. Fifth, the sheer amount and variety of evidence suggests that knowledge is uniquely normatively connected to assertion. I have been unable to observe, either in my own behavior or in others', similar connections between knowledge and any other speech act. But perhaps further investigation will prove me wrong on this point.

What of the assumption that the knowledge rule helps to constitute the practice of assertion? There are only so many candidates for the sort of normativity at issue. Breaking the knowledge rule need not, and typically will not, be either immoral, imprudent, irrational, impolite, or illegal. And, as reviewed in Chapter 3, people's judgments about assertability do not reduce to an evaluation of an assertion's morality, rationality, etiquette, or legality — the "should" in "should assert" transcends those sources of normativity. The only familiar sort of normativity that seems to fit is constitutive. The assumption also coheres with what seems obvious in the following thought experiment.

Imagine a community that speaks a language very similar to our own, except that knowledge-talk is completely absent from the give and take surrounding their declarative utterances. They do not prompt utterances by asking, "Do you know what time it is?" Someone who says, "Pickett's Charge was a serious error," stares in puzzlement upon being asked, "How do you know that?" Questioners are boggled upon hearing "I don't know" in response to their question. Their evaluations of other people's declarative utterances are insensitive to whether the proposition asserted is known or even whether it is true. Upon considering this community, it seems clear to me that they have a completely different information-sharing practice from ours. Assimilating into this community would require unlearning our practice of assertion and learning an entirely new practice. I would feel like a foreigner in their midst. Conversely, teaching them our practice would involve, among other things, sensitizing them to the relevance of knowledge.

In this chapter's final section, I give greater substance to the claim of constitutive normativity, in terms of the evolution and maintenance of communication systems.

Good Enough?

In the previous section, I said that the available evidence supports the assumption that "should" expresses a status more like permission than goodness. But I also acknowledged that further investigation could reveal greater complexity on this point and, furthermore, that nothing in the evidence requires that knowledge is the unique norm of assertion. At the risk of diluting my basic message, I would now like to consider a slightly more complicated view of assertional norms. I am not convinced that this more complicated view is correct, but I discuss it briefly in the spirit of exploration.

You enter some shabby government office to take care of some annoying paperwork for some irritating responsibility. The room is packed. A sign yellowing with age greets you as you enter, "Please take a number and we will be happy to assist you shortly." You take a number, 117, note with disgust that they are presently serving number 9, sit down in one of the cheap, small plastic chairs crammed along the wall, and patiently wait your turn. Hours slowly pass. Finally, your turn approaches. "Now serving number 114," the electronic display reads. You welcome the thought of soon being done with this unpleasant and aggravating experience. But you are also somewhat concerned about the older woman sitting across from you. As difficult as this long wait in cramped, uncomfortable quarters has been for you, it has clearly been more difficult for her. It is essential that she stay and resolve an urgent matter regarding her pension, she tells you, but her aching hip is not making it easy. She continues in earnest, "I see that you have number 117. I have number 122. Would you mind trading numbers with me, please? I could manage either way, of course, but extending me this favor would be a relief." Granting her request would require you to wait in line about an extra ten minutes.

It would be good for you to do her this favor. You should switch numbers. We might praise you for switching numbers. If you did not switch numbers, we might think less of you; we might offer some gentle criticism after the fact; and arguably, you should later regret your decision. All this despite our recognizing that you have a right to your number, that you would be within your rights to refuse her request, and thus that your refusing to switch is morally permissible. In a word, refusing to do the favor in this case is bad but permissible. (For those who do not share the intuition about the case as described, please adjust the case by increasing the older woman's ticket number just enough until you feel it would no longer be wrong for you to refuse.)

A bad but permissible action is sometimes called "suberogatory" (Driver 1992) or an "offence" (Chisholm 1963). This is the inverse of the supererogatory. A supererogatory act is good but not required. The supererogatory is an important normative category that helps us understand our moral judgments about actions in the interval spanning the required and the heroic. Similarly the suberogatory helps us, as one philosopher eloquently put it, to shed light on our normative judgments "lying in the dark corners between right and wrong" (Driver 1992: 295).

If there are suberogatory actions, then why not suberogatory assertions too? If the latter are real, then perhaps belief, or justified belief, sets the standard for a permissible assertion, whereas knowledge sets the standard for a good assertion. On this view, your assertion should express knowledge in the way that you should switch numbers with the elderly woman in the office. We discourage and disapprove of assertions that do not express knowledge, just as we discourage and disapprove of not accommodating the elderly woman.

In earlier work, I suggested that this slightly more complicated account of assertional norms has some virtues (Turri 2014a). In particular, I argued that it could explain much of the observational data discussed above in Chapter 1 without implying that reasonable false assertions are impermissible. When I first conceived of this account, I was responding to repeated objections from professional philosophers who told me, in no uncertain terms, that reasonable false assertions were pretheoretically compelling counterexamples to the knowledge account. I did not find them to be compelling and I thought the familiar distinction between impermissible and blameworthy performances was at least a plausible diagnosis of what might be going on (Williamson 2000: ch. 11; see also DeRose 2002). Nevertheless, mindful of my interlocutors' intelligence and standing, I found the situation somewhat unsettling. I wondered, was I just in the grip of a theory? Was I systematically misinterpreting my social and introspective observations? From a purely theoretical perspective, classifying reasonable false assertions as bad but permissible seemed like a potentially good compromise.

However, the question is not purely theoretical. There is a fact of the matter about how these assertions are ordinarily viewed. When the relevant empirical work was done, the results — especially the results on excuse validation — undermined the initial motivation for postulating the more complicated account. Empirical investigation supplemented the theoretical discussion at a critical juncture, ruling out mistaken objections and thereby avoiding unmotivated amendments and complications. Thus, in purely dialectical terms, no compromise is called for.

Nevertheless, serious inquiry is always about more than dialectic. It is ultimately about learning the truth, uncovering the facts of interest, and the facts do not compromise. Perhaps more complicated accounts deserve consideration for independent reasons. I have discussed one such account here in the hope of setting a constructive example. In particular, I hope it exemplifies how, if the need arises, to rethink the normative relationship between knowledge and assertion without throwing the proverbial baby out with the bathwater.

Super Norm?

We routinely make assertions, form beliefs, and make decisions. These are ubiquitous and unavoidable in the course of ordinary human affairs. It is important to do these things correctly. Our individual and collective well-being often depends on it. Unsurprisingly, then, researchers are keenly interested in what such correctness consists in. While an enormous amount of work has been done on the norms of assertion over the past decade, a related, albeit smaller, body of work has also developed on the norms of belief and decision (or practical reasoning). One intriguing possibility is that knowledge is the norm of all three — assertion, belief, and decision. Is knowledge a super norm?

Requisite Truth

By now the reader is utterly familiar with the view that knowledge is the norm of assertion, the wealth of theoretical and empirical evidence supporting it, and the critic's favorite tactic of producing alleged counterexamples in response. The most common and persistent sort of example features a reasonable false assertion. Interestingly, the exact same tactic is used in response to the hypotheses that knowledge is the norm of belief and of decision.

According to the knowledge account of the norm of belief, you should believe a proposition only if you know that it is true (Williamson 2000: 255-6; see also Sutton 2007; Bach 2008: 77). Much less evidence supports this view about belief than its analog about assertion (for some bits and pieces, see Huemer 2007, 2011; Bird 2007; Turri 2011; Littlejohn 2013). Critics object to the knowledge account of belief by, again, claiming that it is "completely implausible" (McGlynn 2013: 390) and that it faces intuitively compelling counterexamples. Foremost among these are cases of reasonable false belief (Conee 2007; Benton, 2012: 6).

According to the knowledge account of the norm of decision, you should base decisions on a proposition only if you know that it is true (Hawthorne 2004: 29–30; Hawthorne & Stanley 2008; Montminy 2013). Again, less evidence supports this view about decision than its analog about assertion (for some bits and pieces, see Hawthorne & Stanley 2008). And, once again, critics claim that it faces obvious counterexamples and is at odds with our ordinary practice of evaluating decisions. Foremost among the complaints are, yet again, cases of decision based on reasonable false beliefs (Hill & Schechter 2007: 115; Douven 2008: 106–07 n. 9).

All three knowledge accounts — of assertion, belief, and decision — face the same objection. In each case critics object that the account is highly counterintuitive and revisionary. The alleged counterexamples focus on reasonable false beliefs. The implication is that an account that respects ordinary practice will feature a *non-factive* norm. Critics have developed a variety of non-factive views. Perhaps the most popular view is that evidence or justification is the norm of belief and decision.

As discussed in Chapter 3, a series of behavioral experiments showed that critics misdescribed the ordinary way of evaluating reasonable false assertions. This raises the prospect that critics have also overstated their objections to factive accounts of the norms of belief and decision too. Another series of experiments tested this possibility (Turri 2015d). People evaluated beliefs and decisions in cases where a known highly reliable source provides an agent with evidence that a certain proposition was true.

Consider Mario, who manages human resources for a company with thousands of employees. He cannot keep track of all their names by memory, so he maintains a detailed inventory of them. He keeps the inventory up to date. He knows that the inventory is not perfect, but it is extremely accurate. Today his colleague informed him that the immigration office called. If the company employs someone named "Rosanna Winchester," then Mario needs to make an appointment to revise paperwork with immigration, which will take several hours. But if they do not have an employee by that name, then Mario does not need to make an appointment. Mario consults the inventory. It says that he does have an employee by that name. At the end of the story, one group of people was told that the inventory was right. Another group of people was told that the inventory was wrong. Should Mario believe that they employ someone by that name? Should Mario make an appointment with the immigration office?

These results show that critics of factive accounts have mischaracterized the ordinary way of evaluating beliefs and decisions. When the proposition was true, the vast majority of people said that Mario should believe that they employ someone by that name, and that Mario should make an appointment. But when the proposition was false, only a very small minority said that Mario should believe the proposition, or that Mario should make an appointment.

When explaining their mistaken interpretation of cases like Mario's, critics appeal heavily to the fact that the agent has "excellent reasons" or is "epistemically justified" or has sufficient "evidence" (Douven 2008: 106 n. 9; Hill & Schechter 2007: 115; Kvanvig 2009: 145) for the proposition in question. This seems to assume something important about evidence. In particular, it seems to assume that, on any particular occasion, the quality of evidence is insensitive to the truth of the matter. For example, consider Mario's inventory of employee names that he knows to be extremely reliable. When Mario consults the inventory, it says that he has an employee by a certain name. The evidence this provides Mario is equally good regardless of whether the inventory is right on this particular occasion. Or so critics assume.

Something like this view of evidence is popular among contemporary philosophers (e.g. Chisholm 1989: 76; see also BonJour 2003: 185–86; Lehrer & Cohen 1983). As one leading epistemologist puts it, consider a "typical case" where there is "nothing odd" and "things are exactly as the person believes them to be." Compare that to an "unusual case" where "the person has that very same evidence, but the proposition in question is nevertheless false." The "key thing to note" here is that in each case the person "has exactly the same reasons for believing exactly the same thing." Consequently, if the person has a good reason to believe the proposition in either case, then the person has a good reason in both cases (Feldman 2003: 29).

It turns out that this "truth-insensitive" view of evidence is neither natural nor intuitive. The same line of experiments just discussed also investigated people's evaluation of evidence. Changing the truth value of the proposition radically changed how people rated the person's evidence for the proposition. When Mario's inventory was accurate, people overwhelmingly said that his evidence was very good. But when it was inaccurate, very few people said that the evidence was very good. Instead the central tendency was one of ambivalence about the evidence's quality. A related series of experiments showed that this very same pattern of truth-sensitivity emerges for people's judgments about whether a belief is justified, rational, responsible, and reasonable (Turri in press a).

Overall, then, the evidence supports two conclusions. First, it disproves the accusation that factive accounts of belief and decision are counterintuitive or revisionary. The effect of truth on these evaluative judgments was not only statistically significant but also extremely large. From this I conclude belief and decision probably have factive norms. Second, it strongly suggests that critics of factive accounts of the norms of assertion, belief, and decision have been relying on an idiosyncratic view of evidence. In particular, critics have falsely assumed that "it is beyond question at an intuitive level" that the quality of someone's evidence for a proposition is insensitive to the proposition's truth value (BonJour 2003: 186).

Requisite Knowledge

Truth and knowledge are the two leading contenders for a factive norm of belief and decision. The line of research under discussion also investigated the underlying causal relationships among people's evaluations of beliefs, decisions, evidence, knowledge, and truth (Turri 2015d). When controlling for the influence of truth and knowledge judgments, evaluations of evidence (i.e. how good someone's evidence is) did not affect the evaluation of beliefs and decisions (i.e. whether someone should believe or base decisions on a proposition). By contrast, when controlling for the influence of truth and evaluations of evidence, knowledge judgments still deeply affected the evaluation of beliefs and decisions. Crucially, knowledge judgments also mediated truth's effect on these evaluations. This suggests that truth judgments influence these evaluations because they influences knowledge judgments.

A knowledge account of the norms of belief and decision can easily explain these findings. Knowledge requires truth, so the knowledge account predicts that truth will influence the evaluations. But a truth account cannot easily explain the powerful independent influence that knowledge exerts on these evaluations. Neither can a truth account easily explain why knowledge mediates truth's influence. Overall, a knowledge account is a much better fit for the evidence at hand.

Inside and Out

A final set of experiments accentuated the deep normative connection between knowledge and the evaluation of decisions.

It is well documented that people often misunderstand probabilities (e.g. Kahneman & Tversky 1972). But even when they understand the probabilities, interesting differences arise depending on the statistical information's character. For example, mock jurors make different liability judgments across conditions where they assign equal probability to the defendant's guilt (Wells 1992). Suppose that Smith's dog was hit by a bus but no one witnessed the accident. In one version of the case, jurors learn that 80% of the buses operating in the town belonged to the Blue Bus Company, and 20% belonged to the Grey Bus Company. People estimate that it is 80% likely that a Blue bus killed the dog, but they tend to disagree that the jury should find the Blue Bus Company liable. In another version of the case, jurors learn that shortly before the accident, a weigh station attendant made a log entry on the bus that eventually killed the dog. The log says, "Blue bus." It is known that 80% of "Blue bus" entries in the log correctly identify a Blue bus, and 20% incorrectly identify a Grey bus. Again people estimate that it is 80% likely that a Blue bus killed the dog, and they tend to agree that the jury should find the Blue Bus Company liable.

Related to this finding is the "gatecrasher paradox" — if 95% of attendees snuck into the rodeo without paying, then why should the rodeo organizers not be entitled to sue a random attendee for non-payment, or even all attendees (Cohen 1981)? More generally related is the preference for "clinical" to "statistical" decision procedures. People prefer to make decisions based on "observations or impressions" specific to the case at hand rather than empirically established statistics, even when decades of scientific evidence show that relying on the statistics produces better results (Dawes, Faust & Meehl 1989: 1673; see also Dawes 1996; Meehl 1954).

One way to unify these findings draws on the distinction between "inside" and "outside" probabilistic information (Lagnado & Sloman 2004; see also Kahneman & Tversky 1982). "Inside" information concerns a specific item in the case at hand. For example, the weigh station attendant entered information about the very bus that hit the dog. "Outside" information is generic and concerns distributions of properties or patterns. For example, a certain percentage of buses operating in town that day belonged to the Blue Bus Company. The attendant's log book and the distribution of busses each makes it likely that a Blue bus hit the dog, but only the former does so from the "inside." People are less likely to judge that the company is liable based on "outside" evidence.

It has recently been shown that the inside/outside difference also affects knowledge judgments (Friedman & Turri 2015). For instance, suppose that Bob wonders whether his spider plant contains the (fictitious) chemical aracnium. Bob might consult a book on spider plants which reports that 99% of spider plants contain aracnium (outside information), or Bob might conduct a test showing that his spider plant is 99% likely to contain aracnium (inside information). People are more likely to attribute knowledge to Bob when he conducts the test, even though the likelihood is the same in both cases.

Could it be, then, that the inside/outside difference affects judgments about what people should decide to do *because* it affects judgments about what they know? That is, do knowledge judgments mediate the inside/outside effect on decision evaluations? A recent experiment was designed specifically to answer this question, and the answer appears to be "yes" (Turri, Friedman & Keefner in press: Experiment 4). Participants read a story about Gary, who is suing the Blue Cab

Company. Gary's prize-winning rose garden was destroyed by a taxi cab that drove on to his front lawn. During the trial, jurors learned that only two cab companies operate in the town: the Blue Cab Company and the Green Cab Company. At this point, the story ends in one of two ways. One group of people was told that according to a computerized analysis of the video footage from when Gary's garden was destroyed, "80% of cabs on the road were Blue Cabs" (outside information). The other group was told that according to the analysis, "the cab was 80% likely to be a Blue Cab" (inside information). People were then asked to rate the probability that a Blue cab destroyed the garden, whether the jurors know that a Blue cab destroyed the garden, and whether the jurors should rule against the Blue Cab Company.

People in both groups responded to the probability question similarly (on average, they judged the probability to be just below 80%), but they gave significantly different answers to the questions about what the jurors know and should decide. In other words, the inside/outside difference affected knowledge judgments and decision evaluations, but it did not affect probability estimates. Outside information made people significantly less likely to attribute knowledge. It also made them significantly less likely to say that the jurors should rule against the company. Critically, statistical analysis showed that the inside/outside effect on decision evaluations was completely mediated by knowledge judgments. Once we control for the influence of knowledge judgments on decision evaluations, the inside/outside difference has no further effect on decision evaluations. The same pattern was observed using several different examples pertaining to legal and medical decision-making. A follow-up study investigated whether attributions of certainty also mediated the inside/outside effect on decision evaluations (Turri, Friedman & Keefner in press, Experiment 5). Researchers found no evidence that certainty mediated the effect.

Altogether, this suggests that the inside/outside difference affects decision evaluations because it affects knowledge judgments. An easy explanation for this result is that people's decision evaluations are based on their knowledge judgments: they think that what we should do depends on what we know. In a word, knowledge is the norm of decision.

Intuitive Connections

More general theoretical considerations also support knowledge accounts of belief and decision. If, as many researchers suspect, belief is best understood as assertion to oneself (e.g. Sellars 1963: 180; Dummett 1981: 362; see also Adler 2002), then the knowledge account of assertion entails a knowledge account of belief. Alternatively suppose, as many other researchers expect, that belief is prior to assertion in the order of explanation and assertion is best understood as the expression of belief (e.g., Searle 1979; Bach & Harnish 1979). In that case, the knowledge account of belief could help explain why the knowledge account of assertion is true (Bach 2008). Either way, knowledge accounts of belief and assertion form a natural pair. Moreover, for my part, I am suspicious of the idea that one could be adequately positioned to properly believe and assert a proposition while simultaneously being inadequately positioned to make decisions based on it. That is, I find the following combination of claims counterintuitive: with respect to a certain proposition, you should believe it, and you should assert it, but you should not make decisions based on it. Further investigation of the matter is obviously required. But if my suspicion is well founded, then the knowledge account of decision comes along for the ride.

Of course, however theoretically elegant and satisfying such a unified normative pantheon might be, the world is under no obligation to cooperate in such matters. Much more relevant is the empirical evidence discussed above.

A Coincidence?

If it is true that knowledge is the norm of assertion, belief and decision, it seems highly unlikely that this is just a coincidence. If it is not a coincidence, then it is very interesting to ask why these activities all share a common standard. Some have argued that there are serious difficulties in providing a theoretical explanation for why they share a common standard (Brown 2012: 123). But the argument rests on several questionable assumptions. First, it assumes that the norms in question impose perfect requirements (i.e. they are "exceptionless" standards)

(Brown 2012: 129). I am sympathetic to this assumption when it comes to assertion but, as mentioned earlier in this chapter, it should not be treated as a permanently fixed point. Instead we should be open to the possibility that as we deepen and broaden our investigation, earlier assumptions should be revised. Second, the argument assumes that an explanation for a common standard should not depend on "controversial assumptions" (Brown 2012: 130). But here "controversial assumption" seems to mean "goes against assumptions popular in recent analytic epistemology." The track record of recent analytic epistemology, however, is not very good and I am disinclined to show such deference. Third, the argument assumes that we have certain "intuitions about assertion and practical reasoning" (Brown 2012: 139). In particular, it assumes that *how much is at stake* directly affects our intuitive evaluation of assertions and decisions. But recent work on the psychology of these evaluations suggests the opposite conclusion (Turri & Buckwalter in press).

Fourth, and finally, the argument assumes that theories should be measured against alleged intuitions about highly artificial and fanciful thought experiments. For example, one such thought experiment begins by asking us to imagine a subject, Luke, who knows that he lives in a universe ruled by a "whimsical god" who "severely punishes anyone" who bases a decision on a false proposition, even though this god allows false assertions. This creates the possibility, we are told, that Luke should assert a proposition that he should not base decisions on (Brown 2012: 141). Another thought experiment asks us to imagine someone, call her Jessica, setting knives at a dinner table. Jessica believes that there are knives on the table when she is accosted by a skeptic who argues that she has "no evidence" for believing propositions about "the external world," such as the proposition that there are knives on the table. We are then told that Jessica does not seem to be in "a good enough position" to say that there are knives on the table, even though "it seems that she is still in a good enough position" to believe this proposition and base decisions on it (Brown 2012: 140). I find that things seem otherwise to me. But by now we should be highly suspicious of philosophers' claims about what is intuitive in such cases. Equally importantly, it is ill advised to use such peculiar cases to seriously test theories.

Thus, if knowledge is the norm of assertion, belief, and decision, then I submit that we should think that this is no accident, that there is a deeper explanation, and that uncovering it promises to illuminate important properties of assertion, belief, or decision. There seem to be at least two basic approaches to a deeper explanation here. On the first approach, knowledge is most fundamentally the norm of assertion, belief, or decision, and it is the norm of the other two because of this fundamental fact. So, for example, it could be that knowledge is the norm of decision and because of this it is the norm of belief and assertion too. On the second approach, knowledge is the fundamental principle of normative social cognition and, in virtue of this, is applied to the evaluation of all three activities, and likely others too. Testing between these two approaches, and perhaps others, is a worthwhile task for future work.

Why Knowledge?

An assertion should express knowledge because knowledge transmission is the point of the practice of assertion. This is the proximate explanation for why knowledge is the norm of assertion, which I defended at the end of Chapter 1. But other important questions remain unanswered. Why does knowledge play this role? Is it just an accident that knowledge plays this role, or is there something about the nature of knowledge that best suits it for the role? The fact that these questions remain open does not call into question the obvious fact that knowledge is the norm of assertion. Instead, it highlights the need for further investigation. It also presents an exciting opportunity for cross-fertilization between philosophy and the social and life sciences. In an effort to encourage research in this direction, I will offer a hypothesis informed by decades of findings on animal communication.

Even the most rudimentary forms of life, including bacteria, communicate (Crespi 2001: Waters & Bassler 2005; Keller & Surette 2006). Animal communication is rooted in the detection of information about the animal's environment. Cues are detectable properties of conditions that interest the animal such as the presence of a predator, rival, potential mate, or food source. The function of sense organs is to monitor for relevant environmental cues. Communication is based on signals, a special kind of cue whose function is to provide information to another organism for use in making decisions.

Communication is an adaptive behavioral trait shaped by natural selection (Darwin 1872; Lloyd 1971; Otte 1974; Bradbury & Vehrencamp 2011). Communication systems were selected for and evolved because they benefit sender and receiver (Maynard Smith & Harper 2004). Adaptive and stable communication systems make it worth the sender's effort to send signals and worth the receiver's effort to monitor and parse signals. In social species, improved decision making by other group members tends to benefit the sender. For example, social animals, such as the southern mountain cavy or the black-tailed prairie dog, benefit from foraging in larger groups (Taraborelli 2008; Devenport 1989; see also Lima & Bednekoff 1999). In larger groups, individuals can spend less time actively scanning for predators and more time feeding, so helping group members avoid predation benefits alarm callers, and receivers obviously benefit from avoiding predation. Benefit to the sender can be amplified when group membership is positively correlated with kinship (Hamilton 1964; Sherman 1977; Reeve 1997). To take a less obvious example, prey and predator have a mutual interest in sharing certain information. Prey will often stare intently at nearby predators and follow their movements. This signals to the predator that it has lost the element of surprise, which is often enough to call off the hunt (Hasson 1991; FitzGibbon 1994; Zuberbühler, Jenny & Bshary 1999). Prey obviously benefit from not being hunted, and predators benefit from avoiding hunts with a very low probability of success.

When the sender and receiver benefit similarly from how receivers respond to signals, receivers can count on honest signals. But when the preferences of sender and receiver diverge, senders can also benefit from sending dishonest signals. For example, in some species, over two-thirds of predator alarm calls are false and are often merely intended to scare conspecifics away from a preferred food source or to gain mating opportunities (Haftorn 2000; Wheeler 2009; Bro Jørgensen & Pangle 2010; Wheeler & Hammerschmidt 2013). What keeps senders from routinely sending false or misleading signals when it could benefit them? The simple answer is that, in the long run, receivers adapt: they evolve to better detect dishonesty in a signal, ignore certain signals, and attend to more honest signals. Stable and enduring communication systems thus include features that *promote honest signaling*. (In the animal communication literature, these features are often called "honesty guarantees." But this term has potential to mislead because the "guarantees" often simply make honesty likely enough rather than guarantee it.)

Researchers have identified several mechanisms that promote honest signaling. Here I will focus on two of them. The first mechanism is to attend preferentially to "performance signals," which only some signalers can produce (Hurd & Enquist 2005). Some performance signals are indexed to physical characteristics such as body size. For instance, smaller toads cannot croak as deeply as larger toads, so lower-frequency croaks are restricted to larger specimens (Davies & Halliday 1978). Tigers mark territory by scratching tree trunks as high as they can reach, so scratch height is indexed to body size (Thapar 1986; Bradbury & Vehrencamp 2011: 297). Signalers cannot send dishonest signals that their body size prevents them from sending.

Other performance signals are "information constrained" (Hurd & Enquist 2005). For instance, consider the earlier example of pursuit-deterrent signaling by potential prey. An antelope stares at a lioness in the brush and follows her movements, thereby signaling to the lioness "both its alerted state and the futility of continuing the hunt. This signal can be performed only by a signaler who knows the location of the hidden predator" (Hurd & Enquist 2005: 1160). To take another example, neighboring sparrows usually share at least two songs in their repertoire. Sparrows view established neighbors as less threatening than strangers (Temeles 1994). A neighbor's song will be heard frequently, even when the neighbor is respecting the bird's territory. A sparrow typically responds to a neighbor's song by singing a different song it shares with the neighbor ("repertoire matching"), which expresses tolerance. By contrast, a sparrow typically responds to a stranger's song by imitating it ("song matching"), which expresses aggressive intent (Vehrencamp 2001). Since repertoire matching "requires knowledge of the singer's repertoire," or having "committed that bird's repertoire to memory," it is an informationally-constrained signal of neighborhood (Beecher, Campbell, Burt, Hill & Nordby 2000: 22, 25).

The second mechanism is social policing, which involves testing for honesty and retaliating for dishonesty. This is important for signals whose form is arbitrarily associated their significance ("conventional signals"). For example, in some sparrow species the amount of dark plumage on the head and throat is correlated with social dominance (Rohwer 1977). This visible marker of dominance is a "badge of status" that allows conspecifics to settle disputes over resources without resorting to potentially harmful fighting (Dawkins & Krebs 1978). The benefit of a large badge is deference from individuals with small badges. But if a larger badge confers such advantages, then why do subordinates not simply molt into plumage that resembles a higher rank? Because a system of "social control" actively prevents it (Moller 1987). Individuals with large badges frequently challenge one another, ensuring that pretenders will be exposed. In experiments that artificially enlarged the badges of subordinate individuals, the altered individuals suffered "social persecution" and, eventually, bouts of self-imposed "exclusion" from the flock (Rohwer 1977: 116). As two ethologists memorably put it, "This persecution resulted in a nearly fourfold increase in the rate of attacks upon these subordinates and was so severe that several of the dyed birds resorted to visiting the food patch alone" (Rohwer & Rohwer 1978: 1013). Social punishment of dishonest signalers has also been observed in lizards and wasps (Thompson & Moore 1991; Tibbetts & Dale 2004; Tibbets & Izzo 2010).

Behavioral ecologists describe "receiver retaliation" as a "behavioral rule" that disincentivizes dishonest conventional signaling. Indeed, they consider this rule to be "the key factor" that inhibits dishonesty and makes efficient conventional communication systems evolutionarily stable (Bradbury & Vehrencamp 2011: 411). Absent retaliatory costs, dishonesty would proliferate and eventually conventional signals would just be ignored. If the signals are ignored, then senders gain no advantage from producing them and will eventually stop producing them. Conventional communication would be abridged severely, if not abrogated entirely.

Primates retaliate against conspecifics both for providing false information and for withholding relevant information — for being "detected as a cheater" (Hauser 1992: 12137). But retaliation does not always take the form of outright aggression. Sometimes the cost is diminished reputation and distrust from other group members, known as "skeptical responding" (Cheney & Seyfarth 1988; Gouzoules, Gouzoules & Miller 1996). The sophistication of skeptical responding in some monkey species is truly remarkable. For example, a vervet monkey who gives false leopard alarm calls will eventually be ignored on subsequent leopard alarm calls, but this skepticism is *not* transferred to that monkey's eagle alarm calls. Vervets have multiple calls indicating the presence of another group of vervets ("intergroup calls"). The calls are very different acoustically, including a longer "wrr" and a shorter "chutter." A vervet who repeatedly gives false "wrrs" will also have its subsequent "chutters" ignored too. In other words, vervets respond skeptically based on a caller's history as well as a call's abstract semantic properties (Cheney & Seyfarth 1988). Skeptical responding has been observed in other primates and ground squirrels too (Wheeler & Hammerschmidt 2013; Hare & Atkins 2001).

Just like animal communication more generally, human communication is an adaptive trait that benefits sender and receiver. Human communication is subject to the same general evolutionary pressures as any animal communication system. Accordingly, we should expect human communication to exhibit important similarities to animal communication systems. Spoken human language is a paradigm example of conventional communication. An assertion's content is arbitrarily associated with its form and human speech is very cheap to produce. Thus the question arises, what prevents senders from dishonestly asserting to their own advantage? Of course, not infrequently humans do lie and mislead (Feldman, Forrest & Happ 2002; Weiss & Feldman 2006; Vrij 2008). So more specifically the question is, what prevents senders from dishonestly asserting enough to destabilize the practice?

In light of all the observational and behavioral data discussed in the preceding chapters, I propose that the answer is social policing of an information constraint. The information constraint is to have detected or discovered the fact in question. Knowledge just is a true belief manifesting one's powers to detect, discover, or remember a fact (Reid 1764/1997; Greco 2010; Turri 2015b; Turri in press d; Turri in press f). So the information constraint is knowledge, as the researchers quoted above realize. Each part of this proposal has precedent in animal communication more generally. On the one hand, social policing is a mechanism for stabilizing conventional communication systems that has evolved in many species. On the other hand, information-constrained signaling is a standard feature of many animal communication systems. I propose that the human practice of assertion is characterized by a combined application of these two ancient themes: *a socially policed knowledge rule*. Our following this rule stabilizes and sustains the practice of assertion. In this sense, the knowledge rule helps constitute our social practice of assertion.

We test for honesty and accuracy by challenging assertors with questions that explicitly refer to or imply knowledge (see the evidence on challenges in Chapter 1). If the situation calls for it, we will make observations, calculations or consultations of our own to verify the statement. The cost of dishonest or inaccurate assertion is normally a diminished reputation. A dishonest assertion or an outright lie can earn you the label of "liar" and the hostility and disadvantages that go along with it. More serious consequences are possible, including resorting to legal or violent means. A pattern of well-intentioned but ignorant assertions will lower people's estimation of your competence and lead them to trust you less.

Reputational costs need not result from an explicit invocation of the relevant rule. As noted above, monkeys and ground squirrels lose trust in previously inaccurate signalers. But presumably these animals are not conscious of the behavioral rule they are following. Similarly, human infants express surprise when someone falsely identifies an object, just as they express surprise when someone correctly identifies an object that they have not seen (Koenig & Echols 2003; Onishi & Baillargeon 2005; Baillargeon, Scott & He 2010). But, again, presumably human infants are not conscious of the behavioral rule they are following. Even human adults, who can become explicitly aware of the rule, often make normative social judgments implicitly and automatically (Fiske & Taylor 2012: ch. 2).

Although we can follow rules implicitly or unconsciously, like human infants and our animal cousins, we do not always follow rules that way. Three and four year old human children spontaneously keep track of speakers' track records of accuracy and modulate their subsequent decisions and trust of informants accordingly (Koenig, Clement & Harris 2004; Koenig & Harris 2005; Birch, Vauthier & Bloom 2008). By this age children learn better from people to whom they attribute knowledge and their learning is "based on judgments about speakers' knowledge states" (Sabbagh & Baldwin 2001: 1067). By this age children also cite knowledge and ignorance to explain people's verbal performances. When asked why someone was "not good at answering questions," even three year olds say that it is because the person "didn't know" (Koenig & Harris 2005: 1266 ff.). And when asked why they are good at answering questions, children will say, "Because I know." Human adults go beyond even this impressive level of awareness of the content of the rules to an understanding of the conditions that give rise to practices with those very rules. Human practices are extraordinary in part because of our ability to reflect upon, identify, and think critically about the rules that structure them.

We humans are unique in our range of expressive powers. We can communicate with one another about a limitless number of topics, both real and imagined. But this does not imply that the rules which structure and sustain our communicative practices are unique. We face problems similar to those faced by other social species, and the structure of these problems may force natural selection to favor similar solutions repeatedly. When it comes to sustaining communication systems, knowledge is an ancient solution.

Before concluding this chapter, I would like to comment briefly on the conception of knowledge presupposed by this hypothesis. Earlier I defined knowledge as a true belief manifesting one's powers to detect, discover, or remember a fact. However, "belief" might not be the right word here. "Representation" might be better. I say this for three reasons. First, in some cases, people reliably attribute knowledge at higher rates than they attribute belief (Myers-Schulz & Schwitzgebel 2013; Murray, Sytsma & Livengood 2012). Second, in many perfectly ordinary contexts, people's knowledge attributions are not based on belief attributions — the latter do not predict or guide the former (Turri & Buckwalter in press; Turri, Buckwalter & Rose under review). Third, there is abundant evidence that non-human primates attribute knowledge to others, but there is currently no clear evidence that they attribute beliefs or even have the concept of belief (Kaminski, Call & Tomasello 2008; Hare, Call & Tomasello 2001; Flombaum & Santos 2005; Santos, Nissen & Ferrugia 2006; Melis, Call & Tomasello 2006; Marticorena, Ruiz, Mukerji, Goddu & Santos 2011; Martin & Santos 2014; Martin & Santos in press). In light of these three points, it starts to seem unlikely that knowledge requires belief, at least as those categories are understood in ordinary social cognition. Instead, knowledge seems to require a "thinner," more minimal representation of the relevant fact (Buckwalter, Rose & Turri 2015). This conception of knowledge is not unique to humans but instead seems to be part of primate social cognition more generally.

If our concept of knowledge is part of the primate social-cognitive system, then perhaps we can gain further insight into why knowledge, rather than belief or evidence, is so central to normative social cognition in humans. Our normative practices did not emerge out of thin air. They are part of our heritage and so must be based on concepts and categories that our ancestors applied and were sensitive to, at least implicitly. As already mentioned, there is considerable evidence that non-human primates attribute knowledge to others and, moreover, that they use these attributions to guide decision-making. But there is no clear indication that non-human primates ever attribute beliefs. Similarly, there is no indication that non-human primates think of others as having evidence, at least insofar as this is separable from an attribution of knowledge. Accordingly, we might reasonably expect that human normative social cognition would, at least in the first place, rely on attributions of knowledge rather than belief or evidence.

Coda

Benjamin Franklin made it a habit to ask himself every night, "What good have I done today?" I am not in the habit of asking myself this every night, but it seems a good practice upon finishing a book. What good lessons has this book imparted?

The main substantive lesson is, of course, that knowledge is the norm of assertion. With multiple lines of convergent evidence, both theoretical and empirical, proponents of the knowledge account can now assert that their view is true. We now know that it is true — the question is settled. We also have on the table a reasonable, empirically informed, and empirically falsifiable hypothesis about why knowledge is the norm of assertion. In other words, we not only know *that* the knowledge account is true, but we might also now understand *why* it is true. The hypothesis is that knowledge is the norm of assertion because following the knowledge rule stabilizes and sustains the practice of assertion over time within the community. The knowledge rule for assertion is but a new twist on an ancient theme in animal communication systems.

There are some important methodological lessons to be learned. In particular, the research program discussed above can teach us something about progress in philosophy. Progress toward the knowledge account has been cumulative, distributed, responsive, and interdisciplinary. It was cumulative because its brief was built, piece by piece, over time. Interesting isolated observations suggested certain intriguing possibilities, which led to greater alertness to other relevant observations, which eventually reached a critical mass and inspired a concrete hypothesis with empirically testable consequences. The tests were conducted and the hypothesis vindicated, demonstrating that we had alighted upon not only an elegant theory, but also an important fact. It was distributed because it was based on observations and findings from numerous researchers worldwide. It was responsive because it was sharpened in light of criticism and, indeed, some of the most impressive evidence was discovered in response to criticisms. Finally, it was interdisciplinary because it was tested experimentally and placed in a broader context of principles and findings from other disciplines, especially psychology, ethology, behavioral ecology, and evolutionary biology. The broader context was not only consistent with the hypothesis, but it helped to illuminate why the hypothesis is true. All this resembles the process by which modern science advances, producing new knowledge and understanding of the world and our place in it. In the present case, philosophical and scientific enterprises were not only continuous, they were virtually indistinguishable.

Every sensible person believes that the correct answer to philosophical questions must be consistent with scientific knowledge. Many philosophers believe that philosophical inquiry should be closely informed by scientific knowledge. Nevertheless, at least in modern Anglo-American philosophical circles, few believe that philosophical questions are well answered by conducting scientific experiments. Indeed, some believe that it is fundamentally misguided to apply experimental science to answering philosophical questions. But the present work reveals that pessimistic view itself to be deeply misguided. Experimental science turned out to be an excellent way for us to come to know what the norm of assertion is. With respect to this particular philosophical question, in getting beyond speculation, however reasonable and fruitful such speculation may be, philosophical science worked wonders.

References

Achinstein, P. 1983. *The Nature of Explanation* (Oxford: Oxford University Press).

Adler, J.E. 1997. 'Lying, Deceiving, or Falsely Implicating', *The Journal of Philosophy*, 94: 435–52.

— 2002. *Belief's Own Ethics* (Cambridge, MA: MIT Press).

Alicke, M.D., J. Buckingham, E. Zell, & T. Davis. 2008. 'Culpable Control and Counterfactual Reasoning in the Psychology of Blame', *Personality and Social Psychology Bulletin*, 34: 1371–78, http://dx.doi.org/10.1177/0146167208321594

— 1992. 'Culpable Causation', *Journal of Personality and Social Psychology*, 63: 368–78.

— 2000. 'Culpable Control and the Psychology of Blame', *Psychological Bulletin*, 126: 556, http://dx.doi.org/10.1037/0033-2909.126.4.556

— 2008. 'Blaming Badly', *Journal of Cognition and Culture*, 8: 179–86.

Alicke, M.D., & D. Rose. 2010. 'Culpable Control or Moral Concepts?', *Behavioral and Brain Sciences*, 33: 330–31, http://dx.doi.org/10.1017/S0140525X10001664

Aquinas, Thomas. 1273. *Summa Theologica* (Amazon Digital Services).

Aristotle. 1941. 'Posterior Analytics', in Richard McKeon (ed.), trans. by G.R.G. Mure, *The Basic Works of Aristotle* (New York: Random House).

Augustine. 395AD. 'On Lying', in P. Schaff and K. Knight (eds.), *Nicene and Post-Nicene Fathers, First Series* (Christian Literature Publishing), http://www.newadvent.org/fathers/1312.htm

Austin, J.L. 1946. 'Other Minds', *Proceedings of the Aristotelian Society*, 20: 148–87.

— 1956. 'A Plea for Excuses', *Proceedings of the Aristotelian Society*, 57: 1–30.

Ayer, A.J. 1956. *The Problem of Knowledge* (London: Macmillan).

Bach, K. 2008. 'Applying Pragmatics to Epistemology', *Philosophical Issues*, 18: 68–88.

Bach, K., & R.M. Harnish. 1979. *Linguistic Communication and Speech Acts* (Cambridge, MA: MIT Press).

Baillargeon, R., R.M. Scott, & Z. He. 2010. 'False-Belief Understanding in Infants', *Trends in Cognitive Sciences*, 14: 110–18, http://dx.doi.org/10.1016/j.tics.2009.12.006

Bartsch, K., & H.M. Wellman. 1995. *Children Talk About the Mind* (Oxford University Press).

Bauman, R., & J. Sherzer. 1975. 'The Ethnography of Speaking', *Annual Review of Anthropology*, 4: 95–119.

Beecher, M.D., S.E. Campbell, J.M. Burt, C.E. Hill, & J.C. Nordby. 2000. 'Song-Type Matching Between Neighbouring Song Sparrows', *Animal Behaviour*, 59: 21–27, http://dx.doi.org/10.1006/anbe.1999.1276

Benton, M.A. 2012. 'Knowledge Norms: Assertion, Belief, & Action' (Rutgers University, New Brunswick).

Bhatt, R., & R. Pancheva. 2006. 'Conditionals', in M. Everaert, H. Van Riemsdijk, R. Goedemans and B. Hollebrandse (eds.), *The Blackwell Companion to Syntax* (Wiley-Blackwell), pp. 638–86.

Birch, S., S.A. Vauthier, & P. Bloom. 2008. 'Three- and Four-Year-Olds Spontaneously Use Others' Past Performance to Guide Their Learning', *Cognition*, 107: 1018–34, http://dx.doi.org/10.1016/j.cognition.2007.12.008

Bird, A. 2007. 'Justified Judging', *Philosophy and Phenomenological Research*, 74: 81–110, http://dx.doi.org/10.1111/j.1933-1592.2007.00004.x

Black, M. 1952. 'Saying and Disbelieving', *Analysis*, 13: 25–33.

Bok, S. 1978. *Lying: Moral Choice in Public and Private Life* (New York: Pantheon Books).

BonJour, L., & E. Sosa. 2003. *Epistemic Justification: Internalism vs. Externalism, Foundations vs. Virtues* (Malden, MA: Blackwell).

Bradbury, J.W., & S.L. Vehrencamp. 2011. *Principles of Animal Communication*, 2nd edn (Sunderland, Mass: Sinauer Associates).

Bro Jørgensen, J., & W.M. Pangle. 2010. 'Male Topi Antelopes Alarm Snort Deceptively to Retain Females for Mating', *The American Naturalist*, 176: E33–E39, http://dx.doi.org/10.1086/653078

Brown, J. 2008. 'The Knowledge Norm for Assertion', *Philosophical Issues*, 18: 89–103.

— 2012. 'Assertion and Practical Reasoning', *Philosophy and Phenomenological Research*, 84: 123–57.

Buckwalter, W. 2014. 'Factive Verbs and Protagonist Projection', *Episteme*, 11.4: 391–409.

Buckwalter, W., & J. Turri. 2014. 'Telling, Showing and Knowing: A Unified Theory of Pedagogical Norms', *Analysis*, 74: 16–20, http://dx.doi.org/10.1093/analys/ant092

Buckwalter, W., D. Rose, & J. Turri. 2015. 'Belief Through Thick and Thin', *Nous*, 49.4: 748–75, http://dx.doi.org/10.1111/nous.12048

Cheney, D.L., & R.M. Seyfarth. 1988. 'Assessment of Meaning and the Detection of Unreliable Signals by Vervet Monkeys', *Animal Behaviour*.

Chisholm, R., & T.D. Feehan. 1977. 'The Intent to Deceive', *The Journal of Philosophy*, 74: 143–59.

Chisholm, R. 1963. 'Supererogation and Offence: A Conceptual Scheme for Ethics', *Ratio*, 5: 1–14.

— 1966. *Theory of Knowledge*, 1st edn (Englewood Cliffs, NJ: Prentice Hall).

— 1989. *Theory of Knowledge*, 3rd edn (Englewood Cliffs, NJ: Prentice Hall).

Chomsky, N. 1957. *Syntactic Structures* (The Hague: Mouton).

— 1977. *Essays on Form and Interpretation* (Amsterdam: North Holland).

Coffman, E. J. 2014. 'Lenient Accounts of Warranted Assertability', in Clayton Littlejohn and John Turri (eds.), *Epistemic Norms: New Essays on Action, Belief and Assertion* (Oxford University press), pp. 33–59.

Cohen, J. 1981. 'Subjective Probability and the Paradox of the Gatecrasher', *Arizona State Law Journal*: 627–56.

Conee, E. 2007. 'Review of Jonathan Sutton, *Without Justification*', *Notre Dame Philosophical Reviews*, 12, http://ndpr.nd.edu/review.cfm?id=11803

Craig, E. 1990. *Knowledge and the State of Nature: An Essay in Conceptual Synthesis* (Oxford: Oxford University Press).

Crespi, B. J. 2001. 'The Evolution of Social Behavior in Microorganisms', *Trends in Ecology & Evolution*, 16.4: 178-83.

Crockford, C., R.M. Wittig, R. Mundry, & K. Zuberbühler. 2012. 'Wild Chimpanzees Inform Ignorant Group Members of Danger', *Current Biology*, 122.2: 142-46.

Darwin, Charles. 1872. *The Expression of Emotions in Animals and Man* (London: Murray).

Davies, N.B., & T.R. Halliday. 1978. 'Deep Croaks and Fighting Assessment in Toads Bufo Bufo', *Nature*, 274: 683–85, http://dx.doi.org/10.1038/274683a0

Dawes, R., D. Faust, & P. Meehl. 1989. 'Clinical Versus Actuarial Judgment', *Science*, 243: 1668–74, http://dx.doi.org/10.1126/science.2648573

Dawkins, R., & J.R. Krebs. 1978. 'Animal Signals: Information or Manipulation?', in, *Behavioural Ecology: An Evolutionary Approach* (Oxford: Blackwell Scientific), pp. 282–309.

DeRose, K. 2002. 'Assertion, Knowledge, and Context', *The Philosophical Review*, 111: 167–203.

Descartes, René. 2006. 'Meditations on First Philosophy', in Roger Ariew and Donald Cress (eds.), trans. by Roger Ariew and Donald Cress, *Meditations, Objections, and Replies* (Indianapolis: Hackett).

Devenport, J.A. 1989. 'Social Influences on Foraging in Black-Tailed Prairie Dogs', *Journal of Mammalogy*, 70: 166, http://dx.doi.org/10.2307/1381680

Douven, I. 2006. 'Assertion, Knowledge, and Rational Credibility', *Philosophical Review*, 115: 449–85, http://dx.doi.org/10.1215/00318108-2006-010

— 2008. 'Knowledge and Practical Reasoning', *Dialectica*, 62: 101–18, http://dx.doi.org/10.1111/j.1746-8361.2008.01132.x

Driver, J. 1992. 'The Suberogatory', *Australasian Journal of Philosophy*, 70: 286–95, http://dx.doi.org/10.1080/00048409212345181

Dummett, M. 1981. *Frege: Philosophy of Language*, 2nd edn (Cambridge, MA: Harvard University Press).

Fallis, D. 2009. 'What Is Lying?', *Journal of Philosophy*, 106: 29–56.

Feldman, R.S., J.A. Forrest, & B.R. Happ. 2002. 'Self-Presentation and Verbal Deception: Do Self-Presenters Lie More?', *Basic and Applied Social Psychology*, 24: 163–70.

Feldman, R. 2003. *Epistemology* (Upper Saddle River, NJ: Prentice Hall).

Fiske, S.T., & S.E. Taylor. 2012. *Social Cognition: From Brains to Culture*, 2nd edn (Los Angeles: Sage).

FitzGibbon, C.D. 1994. 'The Costs and Benefits of Predator Inspection Behaviour in Thomson's Gazelles', *Behavioral Ecology and Sociobiology*, 34: 139–48, http://dx.doi.org/10.1007/BF00164184

Flombaum, J.I., & L.R. Santos. 2005. 'Rhesus Monkeys Attribute Perceptions to Others', *Current Biology*, 15: 447–52, http://dx.doi.org/10.1016/j.cub.2004.12.076

Frege, G. 1948. 'Sense and Reference', *The Philosophical Review*, 57: 209–30.

Friedman, O., & Turri, J. 2015. 'Is Probabilistic Evidence a Source of Knowledge?' *Cognitive Science*, 39.5: 1062–80. http://doi.org/10.1111/cogs.12182

Gettier, E.L. 1963. 'Is Justified True Belief Knowledge?', *Analysis*, 23: 121–23.

Goldman, A.I. 1976. 'Discrimination and Perceptual Knowledge', *Journal of Philosophy*, 73: 771–91.

— 2009. 'Williamson on Knowledge and Evidence', in Patrick Greenough and Duncan Pritchard (eds.), *Williamson on Knowledge* (Oxford: Oxford University Press), pp. 73–91.

Gouzoules, H., S. Gouzoules, & K. Miller. 1996. 'Skeptical Responding in Rhesus Monkeys (Macaca Mulatta)', *International Journal of Primatology*, 17: 549–68, http://dx.doi.org/10.1007/BF02735191

Grimm, S. R. 2006. 'Is Understanding a Species of Knowledge?', *The British Journal for the Philosophy of Science*, 57: 515–35, http://dx.doi.org/10.1093/bjps/axl015

Grotius, H. 2001. *On the Laws of War and Peace*, A C Campbell (ed.), trans. by A C Campbell (Kitchener, Ontario: Batoche Books).

Guglielmo, S., & B.F. Malle. 2010. 'Can Unintended Side Effects Be Intentional? Resolving a Controversy Over Intentionality and Morality', *Personality and Social Psychology Bulletin*, 36: 1635–47, http://dx.doi.org/10.1177/0146167210386733

Haftorn, S. 2000. 'Contexts and Possible Functions of Alarm Calling in the Willow Tit, Parus Montanus; the Principle of "Better Safe Than Sorry"', *Behaviour*, 137: 437–49.

Hamilton, W.D. 1964. 'The Genetical Evolution of Social Behaviour. I', *Journal of Theoretical Biology*, 7: 1–16, http://dx.doi.org/10.1016/0022-5193(64)90038-4

Hare, B., J. Call, & M. Tomasello. 2001. 'Do Chimpanzees Know What Conspecifics Know?', *Animal Behaviour*, 61: 139–51, http://dx.doi.org/10.1006/anbe.2000.1518

Hare, J., & B. Atkins. 2001. 'The Squirrel That Cried Wolf: Reliability Detection by Juvenile Richardson's Ground Squirrels (Spermophilus Richardsonii)', *Behavioral Ecology and Sociobiology*, 51: 108–12, http://dx.doi.org/10.1007/s002650100414

Hasson, O. 1991. 'Pursuit-Deterrent Signals: Communication Between Prey and Predator', *Trends in Ecology & Evolution*, 6: 325–29, http://dx.doi.org/10.1016/0169-5347(91)90040-5

Hauser, M.D. 1992. 'Costs of Deception: Cheaters Are Punished in Rhesus Monkeys (Macaca Mulatta)', *Proceedings of the National Academy of Sciences*, 89.24: 12137-39.

Hawthorne, J. 2004. *Knowledge and Lotteries* (Oxford: Oxford University Press).

Hawthorne, J., & J. Stanley. 2008. 'Knowledge and Action', *Journal of Philosophy*, 105: 571

Hill, C., & J. Schechter. 2007. 'Hawthorne's Lottery Puzzle and the Nature of Belief', *Philosophical Issues*, 17: 102–22.

Huemer, M. 2007. 'Moore's Paradox and the Norm of Belief', in Susana Nuccetelli and Gary Seay (eds.), *Themes From G.E. Moore: New Essays in Epistemology and Ethics* (Oxford: Oxford University Press).

— 2011. 'The Puzzle of Metacoherence', *Philosophy and Phenomenological Research*, 82: 1–21.

Hurd, P.L., & M. Enquist. 2005. 'A Strategic Taxonomy of Biological Communication', *Animal Behaviour*, 70: 1155–70, http://dx.doi.org/10.1016/j.anbehav.2005.02.014

Kahneman, D., & A. Tversky. 1972. 'Subjective Probability: A Judgment of Representativeness', *Cognitive Psychology*, 3: 430–54.

— 1982. 'Variants of Uncertainty', *Cognition*, 11: 143–57, http://dx.doi.org/10.1016/0010-0277(82)90023-3

Kaminski, J., J. Call, & M. Tomasello. 2008. 'Chimpanzees Know What Others Know, but Not What They Believe', *Cognition*, 109: 224–34.

Keller, L., & M.G. Surette. 2006. 'Communication in Bacteria: An Ecological and Evolutionary Perspective', *Nature Reviews Microbiology*, 4: 249–58, http://dx.doi.org/10.1038/nrmicro1383

Kim, J. 1999. 'Hempel, Explanation, Metaphysics', *Philosophical Studies*, 94: 1–20.

Klein, P. 1981. *Certainty: A Refutation of Scepticism* (Minneapolis: University of Minnesota Press).

Koenig, M.A., & C.H. Echols. 2003. 'Infants' Understanding of False Labeling Events: The Referential Roles of Words and the Speakers Who Use Them', *Cognition*, 87.3: 179-208

Koenig, M.A., & P.L. Harris. 2005. 'Preschoolers Mistrust Ignorant and Inaccurate Speakers', *Child Development*, 76: 1261–77.

Koenig, M.A., F. Clément, & P.L. Harris. 2004. 'Trust in Testimony: Children's Use of True and False Statements', *Psychological Science*, 15: 694–98.

Kvanvig, J. 2008. 'Epistemic Luck', *Philosophy and Phenomenological Research*, 77: 272–81, http://dx.doi.org/10.1111/j.1933-1592.2008.00187.x

— 2009. 'Assertion, Knowledge, and Lotteries', in Duncan Pritchard and Patrick Greenough (eds.), *Williamson on Knowledge* (Oxford: Oxford University Press), pp. 140–60.

— 2003. *The Value of Truth and the Pursuit of Understanding* (Cambridge: Cambridge University Press).

Lackey, J. 2007. 'Norms of Assertion', *Nous*, 41: 594–626.

Lagnado, D., & S.A. Sloman. 2004. 'Inside and Outside Probability Judgments', in D.J. Koehler and Nigel Harvey (eds.), *Blackwell Handbook of Judgment & Decision Making* (Malden, MA: Blackwell), pp. 157–76.

Lehrer, K., & S. Cohen. 1983. 'Justification, Truth, and Coherence', *Synthese*, 55: 191–207.

Lieberman, M.D. 2013. *Social: Why Our Brains Are Wired to Connect* (New York: Crown Publishers).

Lima, S.L., & P.A. Bednekoff. 1999. 'Back to the Basics of Antipredatory Vigilance: Can Nonvigilant Animals Detect Attack?', *Animal Behaviour*, 58: 537–43, http://dx.doi.org/10.1006/anbe.1999.1182

Littlejohn, C. 2013. 'The Russellian Retreat', *Proceedings of the Aristotelian Society*, 113: 293–320, http://dx.doi.org/10.1111/j.1467-9264.2013.00356.x

Lloyd, J.E. 1971. 'Bioluminescent Communication in Insects', *Annual Review of Entomology*, 16: 97–122.

Lorenz, K. 1974. *On Aggression*, trans. by Marjorie Kerr Wilson (San Diego: Harcourt Brace & Company).

MacIver, A.M. 1938. 'Some Questions About 'Know' and "Think"', *Analysis*, 5: 43–50.

Marticorena, D.C.W., A.M. Ruiz, C. Mukerji, A. Goddu, & L.R. Santos. 2011. 'Monkeys Represent Others' Knowledge but Not Their Beliefs', *Developmental Science*, 14: 1406–16, http://dx.doi.org/10.1111/j.1467-7687.2011.01085.x

Martin, A., & L.R. Santos. 2014. 'The Origins of Belief Representation: Monkeys Fail to Automatically Represent Others' Beliefs', *Cognition*, 130: 300–08, http://dx.doi.org/10.1016/j.cognition.2013.11.016

— [n.d.]. 'Origins of Mental State Representations in Nonhuman Primates', *Trends in Cognitive Sciences*.

Matilal, B.K. 1986. *Perception: An Essay on Classical Indian Theories of Knowledge* (Oxford: Oxford University Press).

Maynard Smith, J., & D. Harper. 2004. *Animal Signals* (New York: Oxford University Press).

McDowell, J.H. 1998. *Meaning, Knowledge, and Reality* (Cambridge, MA: Harvard University Press).

McGlynn, A. 2013. 'Believing Things Unknown', *Nous*, 47: 385–407.

Meehl, P.E. 1954. *Clinical vs. Statistical Prediction: A Theoretical Analysis and a Review of the Evidence* (Minneapolis: University of Minnesota Press).

Melis, A.P., J. Call, & M. Tomasello. 2006. 'Chimpanzees Conceal Visual and Auditory Information From Others', *Journal of Comparative Psychology*, 120: 154, http://dx.doi.org/10.1037/0735-7036.120.2.154

Menzel, C. 2012. 'Solving Ecological Problems', in J.C. Mitani, Josep Call, P.M. Kappeler, R.A. Palombit and Joan B. Silk (eds.), *The Evolution of Primate Societies* (Chicago: University of Chicago Press), pp. 609–27.

Milgram, S. 1974. *Obedience to Authority: An Experimental View* (New York: Harper Perennial).

Mill, J.S. 1979. *Utilitarianism*, George Sher (ed.) (Indianapolis: Hackett).

Moller, A.P. 1987. 'Social Control of Deception Among Status Signalling House Sparrows Passer Domesticus', *Behavioral Ecology and Sociobiology*, 20: 307–11, http://dx.doi.org/10.1007/BF00300675

Montminy, M. 2013. 'Why Assertion and Practical Reasoning Must Be Governed by the Same Epistemic Norm', *Pacific Philosophical Quarterly*, 94: 57–68, http://dx.doi.org/10.1111/j.1468-0114.2012.01444.x

Moore, C., D. Bryant, & D. Furrow. 1989. 'Mental Terms and the Development of Certainty', *Child Development*, 60: 167–71.

Moore, G.E. 1912. *Ethics*. Home University Library of Modern Knowledge n. 54 (London: Williams and Norgate, 1912).

— 1959. *Philosophical Papers* (New York: Collier Books).

Murray, D., J. Sytsma, & J. Livengood. 2013. 'God Knows (but Does God Believe?)', *Philosophical Studies*, 166: 83–107, http://dx.doi.org/10.1007/s11098-012-0022-5

Myers-Schulz, B., & E. Schwitzgebel. 2013. 'Knowing That P Without Believing That P', *Nous*, 47: 371–84.

Neta, R., & G. Rohrbaugh. 2004. 'Luminosity and the Safety of Knowledge', *Pacific Philosophical Quarterly*, 85: 396–406.

Nichols, S., & S. Stich. 2003. *Mindreading: An Integrated Account of Pretense, Self-Awareness and Understanding Other Minds* (New York: Oxford University Press).

Noveck, I.A., & A. Reboul. 2008. 'Experimental Pragmatics: A Gricean Turn in the Study of Language', *Trends in Cognitive Sciences*, 12: 425–31, http://dx.doi.org/10.1016/j.tics.2008.07.009

Onishi, K.H., & R. Baillargeon. 2005. 'Do 15-Month-Old Infants Understand False Beliefs?', *Science*, 308: 255–58, http://dx.doi.org/10.1126/science.1106480

Otte, D. 1974. 'Effects and Functions in the Evolution of Signaling Systems', *Annual Review of Ecology and Systematics*, 5: 385–417.

Pagin, P. 2015. 'Problems with Norms of Assertion', *Philosophy and Phenomenological Research*. http://doi.org/10.1111/phpr.12209

Pelling, C. 2011. 'A Self-Referential Paradox for the Truth Account of Assertion', *Analysis*, 71: 688–88, http://dx.doi.org/10.1093/analys/anr093

— 2012. 'Paradox and the Knowledge Account of Assertion', *Erkenntnis*, 78: 977–78, http://dx.doi.org/10.1007/s10670-012-9360-0

Pritchard, D. 2005. *Epistemic Luck* (New York: Oxford University Press).

— 2010. 'Achievements, Luck and Value', *Think*, 9: 19, http://dx.doi.org/10.1017/S1477175610000035

— 2014. 'Epistemic Luck, Safety, and Assertion', in Clayton Littlejohn & John Turri (eds.), *Epistemic Norms: New Essays on Action, Belief and Assertion* (Oxford: Oxford University Press).

Psillos, S. 2002. *Causation & Explanation* (Montreal: McGill-Queen's).

Reeve, H.K. 1997. 'Evolutionarily Stable Communication Between Kin: A General Model', *Proceedings of the Royal Society B: Biological Sciences*, 264: 1037–40.

Reid, T. 1997. *An Inquiry into the Human Mind on the Principles of Common Sense*, Derek R. Brookes (ed.) (University Park, PA: Pennsylvania State University Press).

Rescorla, M. 2009. 'Assertion and Its Constitutive Norms', *Philosophy and Phenomenological Research*, 79: 98–130.

Reynolds, S.L. 2002. 'Testimony, Knowledge, and Epistemic Goals', *Philosophical Studies*, 110: 139–61.

Rohwer, S. 1977. 'Status Signalling in Harris Sparrows: Some Experiments in Deception', *Behaviour*, 26.4: 1012.

Rohwer, S., & F.C. Rohwer. 1978. 'Status Signalling in Harris Sparrows: Experimental Deceptions Achieved', *Animal Behaviour*, 26: 1012–22, http://dx.doi.org/10.1016/0003-3472(78)90090-8

Rose, D., W. Buckwalter, & J. Turri. 2014. 'When Words Speak Louder Than Actions: Delusion, Belief and the Power of Assertion', *Australasian Journal of Philosophy*, 92: 683–700, http://dx.doi.org/10.1080/00048402.2014.909859

Ross, L., & R.E. Nisbett. 2011. *The Person and the Situation: Perspectives of Social Psychology* (London: Pinter & Martin).

Roth, D., & A.M. Leslie. 1991. 'The Recognition of Attitude Conveyed by Utterance: A Study of Preschool and Autistic Children', *British Journal of Developmental Psychology*, 9: 315–30, http://dx.doi.org/10.1111/j.2044-835X.1991.tb00880.x

Sabbagh, M.A., & D.A. Baldwin. 2001. 'Learning Words From Knowledgeable Versus Ignorant Speakers: Links Between Preschoolers' Theory of Mind and Semantic Development', *Child Development*, 72: 1054–70, http://dx.doi.org/10.1111/1467-8624.00334

Santos, L.R., A.G. Nissen, & J.A. Ferrugia. 2006. 'Rhesus Monkeys, Macaca Mulatta, Know What Others Can and Cannot Hear', *Animal Behaviour*, 71: 1175–81, http://dx.doi.org/10.1016/j.anbehav.2005.10.007

Searle, J.R. 1979. *Expression and Meaning* (Cambridge: Cambridge University Press).

— 2001. *Rationality in Action* (Cambridge, MA: MIT Press).

Sellars, W. 1963. *Science, Perception and Reality* (Atascadero, CA: Ridgeview Publishing Company).

— 1975. 'Epistemic Principles', in H.N. Castaneda (ed.), *Action, Knowledge, and Reality* (Indianapolis: Bobbs-Merrill).

Sextus Empiricus. [n.d.]. *Outlines of Pyrrhonism.*

Shakespeare, W. 1607. *Antony and Cleopatra,* Eric M. Johnson (ed.), http://opensourceshakespeare.org/views/plays/playmenu.php?WorkID=antonycleo, http://www.opensourceshakespeare.org/views/plays/playmenu.php?WorkID=antonycleo

Shatz, M., H.M. Wellman, & S. Silber. 1983. 'The Acquisition of Mental Verbs: A Systematic Investigation of the First Reference to Mental State', *Cognition*, 14: 301–21, http://dx.doi.org/10.1016/0010-0277(83)90008-2

Sherman, P.W. 1977. 'Nepotism and the Evolution of Alarm Calls', *Science*, 197: 1246-53.

Slote, M. 1979. 'Assertion and Belief', in Dancy Jonathan (ed.), *Papers on Language and Logic* (Keele: Keele University Library). Reprinted in Slote, M. (2010), *Selected Essays* (New York: Oxford University Press).

Smithies, D. 2012. 'The Normative Role of Knowledge', *Nous*, 46: 265–88, http://dx.doi.org/10.1111/j.1468-0068.2010.00787.x

Sodian, B., & H. Wimmer. 1987. 'Children's Understanding of Inference as a Source of Knowledge', *Child Development*, 58: 424–33.

Sosa, E. 1991. *Knowledge in Perspective* (Cambridge: Cambridge University Press).

Sperber, D., & I.A. Noveck. 2004. 'Introduction', in D. Sperber & I.A. Noveck (eds.), *Experimental Pragmatics* (New York: Palgrave Macmillan), pp. 1–22.

Stanley, J. 2008. 'Knowledge and Certainty', *Philosophical Issues*, 18: 35–58.

Starmans, C., & O. Friedman. 2012. 'The Folk Conception of Knowledge', *Cognition*, 124: 272–83, http://dx.doi.org/10.1016/j.cognition.2012.05.017

Sutton, J. 2007. *Without Justification* (Cambridge, MA: MIT Press).

Taraborelli, P. 2008. 'Vigilance and Foraging Behaviour in a Social Desert Rodent, Microvacia Australis (Rodentia Caviidae)', *Ethology Ecology & Evolution*, 20: 245–56.

Temeles, E.J. 1994. 'The Role of Neighbours in Territorial Systems: When Are They 'Dear Enemies'?', *Animal Behaviour*, 47: 339–50, http://dx.doi.org/10.1006/anbe.1994.1047

Thapar, V. 1986. *Tiger: Portrait of a Predator* (London: Collins).

Thompson, C.W., & M.C. Moore. 1991. 'Throat Colour Reliably Signals Status in Male Tree Lizards, Urosaurus Ornatus', *Animal Behaviour*, 42: 745–53, http://dx.doi.org/10.1016/S0003-3472(05)80120-4

Tibbetts, E.A., & A. Izzo. 2010. 'Social Punishment of Dishonest Signalers Caused by Mismatch Between Signal and Behavior', *Current Biology*, 20: 1637–40, http://dx.doi.org/10.1016/j.cub.2010.07.042

Tibbetts, E.A., & J. Dale. 2007. 'Individual Recognition: It Is Good to Be Different', *Trends in Ecology & Evolution*, 22: 529–37, http://dx.doi.org/10.1016/j.tree.2007.09.001

Tinbergen, N. 1963. 'On Aims and Methods of Ethology', *Zeitschrift Für Tierpsychologie*, 20: 410–33, http://dx.doi.org/10.1111/j.1439-0310.1963.tb01161.x

Turri, J. 2010a. 'Epistemic Invariantism and Speech Act Contextualism', *Philosophical Review*, 119: 77–95, http://dx.doi.org/10.1215/00318108-2009-026

— 2010b. 'Prompting Challenges', *Analysis*, 70: 456–62, http://dx.doi.org/10.1093/analys/anq027

—2011. 'The Express Knowledge Account of Assertion', *Australasian Journal of Philosophy*, 89: 37–45, http://dx.doi.org/10.1080/00048401003660333

— 2012a. 'In Gettier's Wake', in Stephen Hetherington (ed.), *Epistemology: The Key Thinkers* (London: Continuum), pp. 214–29.

— 2012b. 'Preempting Paradox', *Logos & Episteme*, 3: 659–62.

— 2012c. 'Pyrrhonian Skepticism Meets Speech-Act Theory', *International Journal for the Study of Skepticism*, 2: 83–98, http://dx.doi.org/10.1163/221057011X588037

— 2013a. 'Knowledge Guaranteed', *Nous*, 47: 602–12.

— 2013b. 'The Test of Truth: An Experimental Investigation of the Norm of Assertion', *Cognition*, 129: 279–91, http://dx.doi.org/10.1016/j.cognition.2013.06.012

—2014a. 'Knowledge and Suberogatory Assertion', *Philosophical Studies*, 167: 557–67, http://dx.doi.org/10.1007/s11098-013-0112-z

— 2014b. 'The Problem of ESEE Knowledge', *Ergo*, 1: 101–27.

— 2014c. 'You Gotta Believe', in Clayton Littlejohn and John Turri (eds.), *Epistemic Norms: New Essays on Action, Belief and Assertion* (Oxford: Oxford University Press), pp. 193–99.

— 2015a. 'Evidence of Factive Norms of Belief and Decision', *Synthese*: 1–22, http://dx.doi.org/10.1007/s11229-015-0727-z

— 2015b. 'From Virtue Epistemology to Abilism: Theoretical and Empirical Developments', in Christian Basil Miller, Michael R Furr, Angela Knobel and William Fleeson (eds.), *Character: New Directions from Philosophy, Psychology, and Theology* (New York: Oxford University Press).

— 2015c. 'Selfless Assertions: Some Empirical Evidence', *Synthese*, 192.4: 1221–33. http://doi.org/10.1007/s11229-014-0621-0

— 2015d. 'Evidence of Factive Norms of Belief and Decision. *Synthese*, 192.12: 4009–30. http://doi.org/10.1007/s11229-015-0727-z

— 2015e. 'Knowledge and the Norm of Assertion: A Simple Test', *Synthese*, 192.2: 385–92. http://doi.org/10.1007/s11229-014-0573-4

— 2016. 'Knowledge Judgments in "Gettier" Cases', in J. Sytsma & W. Buckwalter (eds.), *A Companion to Experimental Philosophy* (Malden, MA: Wiley-Blackwell).

— In press a. 'The Radicalism of Truth-Insensitive Epistemology: Truth's Profound Effect on the Evaluation of Belief', *Philosophy and Phenomenological Research* http://dx.doi.org/10.1111/phpr.12218

— In press b. 'Knowledge and Assertion in "Gettier" Cases', *Philosophical Psychology*.

— In press c. 'Knowledge, Certainty and Assertion', *Philosophical Psychology*.

— In press d. 'A New Paradigm for Epistemology: From Reliabilism to Abilism', *Ergo*.

— In press e. 'Vision, Knowledge, and Assertion', *Consciousness and Cognition*.

— In press f. 'Epistemic Situationism and Cognitive Ability', in Mark Alfano and Abrol Fairweather (eds.), *Epistemic Situationism* (Oxford: Oxford University Press).

— In press g. 'The Point of Assertion is to Transmit Knowledge', University of Waterloo.

— Under review. 'The Distinctive "Should" of Assertability', University of Waterloo.

Turri, J., & P. Blouw. 2015. 'Excuse Validation: A Study in Rule-Breaking', *Philosophical Studies*, 172: 615–34, http://dx.doi.org/10.1007/s11098-014-0322-z

Turri, J., & W. Buckwalter. In press. 'Descartes's Schism, Locke's Reunion: Completing the Pragmatic Turn in Epistemology', *American Philosophical Quarterly*.

Turri, J., Buckwalter, W., & Blouw, P. 2015. 'Knowledge and Luck', *Psychonomic Bulletin & Review*, 22(2), 378–90. http://doi.org/10.3758/s13423-014-0683-5

Turri, J., W. Buckwalter, & D. Rose. Under review. 'Actionability Judgments Cause Knowledge Judgments', University of Waterloo.

Turri, J., O. Friedman, & A. Keefner. In press. 'Knowledge Central: A Central Role for Knowledge Attributions in Social Evaluations', *Quarterly Journal of Experimental Psychology*.

Turri, A., & Turri, J. 2015. 'The Truth about Lying', *Cognition*, 138, 161–68. http://doi.org/10.1016/j.cognition.2015.01.007

Turri, A., & Turri, J. Under review. 'Lying, Assertion, Uptake, and Intent', University of Waterloo.

Unger, P. 1975. *Ignorance: A Case for Skepticism* (Oxford: Clarendon Press).

Vehrencamp, S.L. 2001. 'Is Song-Type Matching a Conventional Signal of Aggressive Intentions?', *Proceedings of the Royal Society B: Biological Sciences*, 268: 1637–42, http://dx.doi.org/10.1098/rspb.2001.1714

Vrij, A. 2008. *Detecting Lies and Deceit: Pitfalls and Opportunities*, 2nd edn (Hoboken, NJ: John Wiley & Sons).

Waters, C.M., & B.L. Bassler. 2005. 'Quorum Sensing: Cell-to-Cell Communication in Bacteria', *Annual Review of Cell and Developmental Biology*, 21: 319–46.

Weiss, B., & R.S. Feldman. 2006. 'Looking Good and Lying to Do It: Deception as an Impression Management Strategy in Job Interviews', *Journal of Applied Social Psychology*, 36: 1070–86.

Wells, G.L. 1992. 'Naked Statistical Evidence of Liability: Is Subjective Probability Enough?', *Journal of Personality and Social Psychology*, 62: 739, http://dx.doi.org/10.1037/0022-3514.62.5.739

Wheeler, B.C. 2009. 'Monkeys Crying Wolf? Tufted Capuchin Monkeys Use Anti-Predator Calls to Usurp Resources From Conspecifics', *Proceedings of the Royal Society B: Biological Sciences*, 276: 3013–18, http://dx.doi.org/10.1073/pnas.0706741104

Wheeler, B.C., & K. Hammerschmidt. 2013. 'Proximate Factors Underpinning Receiver Responses to Deceptive False Alarm Calls in Wild Tufted Capuchin Monkeys: Is It Counterdeception?', James P. Higham & Stuart Semple (eds.), *American Journal of Primatology*, 75: 715–25, http://dx.doi.org/10.1002/ajp.22097

Williams, B. 2002. *Truth and Truthfulness* (Princeton: Princeton University Press).

Williamson, T. 1996. 'Knowing and Asserting', *Philosophical Review*, 105: 489–523.

— 2000. *Knowledge and Its Limits* (Oxford: Oxford University Press).

Wittgenstein, L. 1975. *On Certainty*, G.E.M. Anscombe & G.H. von Wright (eds.), trans. by Denis Paul & G.E.M. Anscombe (Malden, MA: Blackwell).

Wright, S. 2014. 'The Dual-Aspect Norms of Belief and Assertion', in Clayton Littlejohn & John Turri (eds.), *Epistemic Norms: New Essays on Action, Belief and Assertion* (Oxford: Oxford University Press), pp. 239–58.

Zuberbühler, K., D. Jenny, & R. Bshary. 1999. 'The Predator Deterrence Function of Primate Alarm Calls', *Ethology*, 105: 477–90, http://dx.doi.org/10.1046/j.1439-0310.1999.00396.x

Index

abstentions 8
accuracy 8, 46, 47, 56, 82, 83
accusations 9, 24, 33
activities 46, 75, 77
adaptive behavioral traits 78, 81
affirmation 26
aggressiveness 9, 22, 23, 24, 33, 80
agreement ratings 15, 16, 54, 58, 72
analytic epistemology 76
ancient skeptics 2
animal
  behavior 4
  communication 5, 77
apparent evidence 42
Aristotle 4, 31
assertability judgments 12, 14, 15, 57, 64
assertion under oath 26
assurance 25
audiocentric responses. *See* listener-centered responses
Augustine 35
authority 8, 9, 10, 20, 24, 27, 33, 62

background assumptions 27
behavioral ecology 88
belief attributions 55, 84
biconditionals 50

birds 79, 80
blameless rule-breaking. *See* excuse validation
blame validation 45

calculation 28
case structure 41, 44
causal modeling 37
causal structure 44
certainty 58
challenges 2, 8, 9, 20, 24, 33, 49, 56, 64
cheap barn cases 43
cheating 34, 81
child psychology. *See* developmental psychology
cognitive psychology 11
commands 52
commitments 26, 28
communicative intent 27
competence 12, 35, 43, 82
complexity 5
concurrent restrictions 62, 64
conjunctions 9
constitutive norms 62, 64, 65
context 8, 23, 25, 52
continuity of philosophy and science 3, 88
contradictions 51, 52

controversial assumptions 76
conventional
  mechanisms 26, 27
  signals 80
conventions 26, 27, 81
counterexamples 51, 56, 61, 67, 68, 69
creationism 54
credibility 29
criticism 66
cues 78

data 4. *See* experimental data, observational data, pre-theoretic data
deception. *See* lying
decision evaluations 73, 74
decisions 68, 69, 71, 72, 75, 76, 77
declarative sentence 2
definite propositions 52
demonstration. *See* instruction
denial 27, 28
deontological rule 62
developmental psychology 10, 55, 83
disapproval 35, 66
discretion 62, 63
disunified responses 57
doubt 53, 55

empirical research 3
entitlement 27
environmental threat 41
epistemic
  state 8
  status 57
escalation 9
ethogram 4
ethology 4, 80, 88
evaluative judgments 40, 50, 57, 64, 69, 71, 72, 74
evidence 39, 42, 70, 71, 73

evolutionary
  biology 88
  theory 54
exceptionless standards 75
exceptions 22, 32, 57
excuse validation 47
experimental
  controls 11, 43, 71, 74
  data 11, 40, 43, 54, 57, 58, 69, 70, 72, 73
  research 4
  science 88
explanation 16, 31, 32, 33, 34, 42, 50, 75, 76, 77
explanatory disconnect 42, 43
express knowledge account of assertion 17

factive norms 12, 13, 69, 70, 71
facts 22, 31
failure 20, 36
fairness 46
fake barn cases 41, 42, 43
fallible evidence 42
false assertions 50
Franklin, Benjamin 87

general conversational mechanisms 27
Gettier
  cases 41, 43, 44
  Edmund 41
Gettiered
  assertions 41
  beliefs 41
group membership 78, 81
guesses 25, 52

hedging 10, 24
Hempel, Carl 34
honest signals. *See* signaling

honesty guarantees 79
human
  civilization 21
  condition 61

ignorance 21, 30, 83
ignorant assertions 39, 41, 82
implicit judgments 43, 50, 83
inconsistency 9, 10, 24, 33, 51
inducing belief 20
inference 28
information 21, 73, 77, 78, 81
  constraints 79, 80, 82
  inside vs outside 73, 74
  sharing practices 61, 65
  transmission of 21, 22, 25
instruction 22, 23, 24, 25
intellectual prejudice 5
intergroup calls 81
introspective observation 7, 11
intuition pump 13, 56
isolated utterances 50, 52

justification 10, 29, 56, 69

kinship 78
know how 25
knowledge
  attributions. *See* psychology of
    knowledge attributions
  judgments 13, 14
  propositional vs procedural 22, 31
  second-order vs first-order 28, 29
  transmission 17, 19, 20, 77
knowledge account of assertion. *See*
  simple knowledge account of
  assertion
known-false account of lying. *See*
  lying

legality 9, 33, 74
lie-detection 35
linguistic
  anthropology 11
  community 11
  data 3
  performance 31
  production and interpretation 11
  rules 11
listener-centered responses 20
locked doors 25, 26, 28
luck 39, 41
lying 37

matched conditions 43, 47
mechanisms 26, 27, 28, 79, 80, 82
memory 39, 40, 69, 80
mental state
  attributions 55
  verbs 10
mercy 22, 27, 32
Mill, John Stuart 47
misunderstanding 11, 22, 32, 72
modulating assertion 10
moral psychology 5

naturalistic observation 4
natural selection 78, 83
negative assertions 40
non-factive norms 69
non-human
  animals 2
  primates 84
norm
  of assertion 2, 7, 12, 30, 51, 52, 56,
    61, 62, 63, 65, 75, 77, 87
  of belief 68, 69, 71, 75, 77
  of decision 68, 69, 71, 74, 75, 77
  of explanation 31, 32, 33

of guaranteeing 28
of instruction 25
of questioning. *See* questioning
of showing. *See* showing
normative
  conversational mechanisms 27, 29
  intuitions 11, 12

observational data 7, 11, 49
ordinary
  competence. *See* competence
  judgments 41
  speech 8
organisms 78
organizing principles 2

paradox 50, 51, 72
pedagogy 22
perceptual evidence 42
perfect vs imperfect requirements 62, 63, 64, 75
performance errors 13, 55
permission 8, 62, 63, 66, 67
philosophy of science 34
phonocentric responses. *See* speaker-centered responses
Plato 29
politeness 27
possibility of error 42
potential 26, 27, 29
powers 82, 83, 84
practical reasoning 68, 76
pragmatics 11, 35
praise 66
predators 79
prefacing 18, 19
pre-theoretic data 49, 50
prime pedagogical principle 25
probability 72, 73, 74
progress in philosophy 61, 87

promising 12, 26
prompts 7, 18, 22, 56, 57
propositions 2
prudence 30, 63
psycholinguistic research 11
psychology 11, 55, 76, 88
  of knowledge attributions 44, 85

questioning 18, 30
  implicitly vs explicitly 9

rationality 55, 64, 71
reasonable false assertions 39, 40, 44, 45, 47, 63, 67, 68, 69
receiver retaliation 80
regression analysis 13, 37, 48
relevance conditionals 19
religious
  belief 54, 55
  texts 3
repertoire matching 79, 80
representation 84
reputational costs 81, 82
requests 21, 32
responsibility 26
rule-breaking 47
rule-governed activity 3

safety 54
saying that you know 26
selfless assertions 53, 54, 55
self-representation 8, 9, 10, 24, 33, 56
semantics 11
Shakespeare, William 3, 4
showing 22, 23, 24, 25
signaling 2, 81
simple knowledge account of assertion 3, 17
skeptical responding 81
skill transmission 21, 22, 25

social
  cognition 40, 55, 58, 84
  controversy 54
  dominance 80
  observation 7, 11
  policing 80, 82
  practice 2, 5, 51, 62
  psychology 35
  status 55
song matching 80
speaker-centered responses 20
speech acts 12, 52, 62, 63
stakes 13, 40, 76
statements 35, 40, 45, 46
statistical
  analysis 74
  decision procedures 73
stimuli 15, 40
suberogatory action 66
supererogatory action 66
sustaining practices 83
swearing 26
syntax 11

telling 22
testable predictions 11, 12
theoretical neutrality 41
theory-laden intuitions 50
thought experiments 35, 53, 55, 76
truth sensitivity 70, 71

understanding 34
unified responses 57
unintentional rule-breaking. *See* excuse validation
unlucky falsehoods 41
unwritten rules 4

vaccines 54
value of knowledge 30
vervet monkeys 81
vindication 9

witnesses 26, 72
Wittgenstein, Ludwig 28
written assertions 13

This book need not end here...

At Open Book Publishers, we are changing the nature of the traditional academic book. The title you have just read will not be left on a library shelf, but will be accessed online by hundreds of readers each month across the globe. We make all our books free to read online so that students, researchers and members of the public who can't afford a printed edition will have access to the same ideas.

Customise

Personalise your copy of this book or design new books using OBP and third-party material. Take chapters or whole books from our published list and make a special edition, a new anthology or an illuminating coursepack. Each customised edition will be produced as a paperback and a downloadable PDF. Find out more at https://www.openbookpublishers.com

Donate

If you enjoyed this book, and feel that research like this should be available to all readers, regardless of their income, please think about donating to us. We do not operate for profit and all donations, as with all other revenue we generate, will be used to finance new Open Access publications.

https://www.openbookpublishers.com/isbn/9781783741830

For further information, please visit our website at
https://www.openbookpublishers.com

Knowledge is for sharing

You may also be interested in:

**Metaethics from a First Person Standpoint
An Introduction to Moral Philosophy**
Catherine Wilson

https://www.openbookpublishers.com/product/317

**Foundations for Moral Relativism:
Second Expanded Edition**
J. David Velleman

https://www.openbookpublishers.com/product/416

Beyond Price: Essays on Birth and Death
J. David Velleman

https://www.openbookpublishers.com/product/349

www.ingramcontent.com/pod-product-compliance
Lightning Source LLC
Chambersburg PA
CBHW071220160426
43196CB00012B/2355